Leisure in Contemporary Society

K. ROBERTS

Department of Sociology, Social Policy and Social Work Studies
University of Liverpool, UK

CABI *Publishing*

CABI *Publishing* is a division of CAB *International*

CABI Publishing
CAB International
Wallingford
Oxon OX10 8DE
UK

Tel: +44 (0)1491 832111
Fax: +44 (0)1491 833508
Email: cabi@cabi.org
Web site: www.cabi.org

CABI Publishing
10 E 40th Street
Suite 3203
New York, NY 10016
USA

Tel: +1 212 481 7018
Fax: +1 212 686 7993
Email: cabi-nao@cabi.org

A catalogue record for this book is available from the British Library, London, UK.

Library of Congress Cataloging-in-Publication Data
Roberts, Kenneth, 1940-
Leisure in contemporary society/K. Roberts.
p. cm.
Includes bibliographical references (p.) and index.
ISBN 0-85199-338-9 (alk. paper)
1. Leisure--Sociological aspects. 2. Recreation--Sociological aspects. I. Title.
GV14.45.R63 1999
306.4′812--dc21 99-22426
CIP

ISBN 0 85199 338 9

First printed 1999
Reprinted 2001

Typeset by York House Typographic
Printed and bound in the UK by Biddles Ltd, Guildford and King's Lynn

Contents

Tables and Boxes

Tables

Boxes

Preface

There is less need than when I first began to write on the subject to justify taking leisure seriously. Sceptics have been defeated by history if not by the weight of argument. Leisure is now rampant. In the West it has been the main 20th century force changing people's everyday lives. The World Wars were more dramatic but their impact was shorter term. The containment of working time and the rise in spending power have been long-term trends, and are likely to continue at least through the early decades of the 21st century. Beyond that, who knows? The growth of leisure has not eradicated or even, in itself, lessened social class, gender or age differences, but it has transformed how these differences are expressed, challenged, modified or reproduced from year to year and from generation to generation.

Momentous change, actual or alleged, has always been sociology's life blood. In earlier times we debated the likelihood of a proletarian revolution, then the impact of the welfare state, then the significance of working class affluence. Attention then switched to the genderquake, consumer culture, post-industrialism and post-Fordism. Nowadays sociology is preoccupied with whether our society is becoming late modern or even postmodern. These debates have made the study of leisure fashionable rather than fringe. Sometimes the subject matter is called consumption. The intellectual arena may be cultural studies rather than leisure studies. The key issue may be defined as people's quests for alternative sites to construct identities, as older divisions and structures become fragile. A plain fact is that leisure research and theory are inevitably embroiled in, and have much to contribute, to all these debates. Leisure theory and research offer, first, a longer historical perspective and, second, a robust body of knowledge on how people's everyday lives interface with the organization of work, domestic roles and the life course. Equally important today, leisure research actually has the evidence to test rather than simply make claims about, or debate the significance of, present-day consumer cultures and the related lifestyles. Leisure is important. Leisure research is now mainstream. It is no longer a backwater into which only specialists need venture.

This book is a product of years of engagement with leisure research and theory, and participation in debates about the character of modern societies, changes therein, and how these impact on various groups. I have enjoyed every minute and am grateful for the mostly friendly, but always productive cut and thrust. I accept sole and total responsibility for everything in the following pages, but none of my output in recent years would have reached a wider audience without the assistance of the wonderful Clare Minghella who word-processed the successive drafts of this manuscript.

K. Roberts
Liverpool University
December 1998

Chapter 1

Leisure: Past and Present

Modern Leisure

This book is about leisure in present-day Western societies with market economies, democratic political systems, and written or unwritten constitutions that protect citizens' rights from encroachment by the state. The leisure in our lives is a product of the modern organization of work, our market economies, the civil liberties that we enjoy, and the weakening of the family, community and religious controls that prescribed and enforced common ways of life in earlier times.

There are plenty of comparisons with other types of societies in the following chapters, and the above definition is not meant to imply that other peoples had, and still have, no leisure; only that a rather different concept is needed to convey the main features. Sometimes the balance of advantage lies with definitions that can be applied in all times and places. In the case of leisure, however, it seems important to acknowledge by definition that, for example under communism, people's lives were different not only in their having different amounts of leisure, or using it in different ways than in Western countries, but also in that what they were using was different. Pre-modern societies had practices that, in some ways, were the functional equivalents, with similar causes and consequences, to our kind of leisure. It is possible to define leisure in ways that can be mapped onto pre-modern ways of life, but doing this obscures the fact that the people concerned lived and understood their own lives and circumstances without coining such a concept. It was only towards the end of the 19th century that the term leisure began to be applied to the lives of Britain's industrial workers (Cunningham, 1980). It is no accident that this was when Britain had become the first modern industrial society.

Leisure is highly context dependent. This will become clear in the following chapters. Even within Western societies it can be argued that men and women, and the employed, unemployed and retired for example, do not share one common type of leisure. What leisure is can vary from group to group within a country depending on each group's circumstances, but the variations between societies, certainly between types of civilizations, are even more profound. Within Western countries, despite the internal variations, it is possible to identify a dominant form of leisure which is produced by four contextual features (the organization of work, the market economy, liberal democracy and the decline of the community).

The organization of work

Our leisure is a product, first and foremost, of the modern organization of work. Some work has not been modernized; housework is an example. But in all modern societies work is ordinarily taken to mean paid work, and most paid work is modernized, meaning here that it is compartmentalized and rationalized.

Paid work is usually done at specific times, at workplaces, and under work-specific authority. This is what is meant by compartmentalization. Work thereby becomes a part of life rather than embedded in multi-functional groups such as families. With the development of modern industries work was taken from its former family and community contexts and located in offices and factories. In these modern work organizations, business is organized by the clock. People do not work when nature decrees that things can or have to be done. There are specified hours of work. The clock dictates when each working day begins and ends. While at work most people are subject to the orders of bosses whose authority stops at the factory or office doors and at the end of each working day. This is different from the relationships that prevailed between lords and serfs in feudal times.

Modern work is compartmentalized and also rationalized. Business is organized so that things are done efficiently. Work is not governed by tradition. If a new machine will perform more effectively then custom and practice are discarded. This does not mean that work has to be disagreeable, but when people like their jobs this is a fortuitous by-product rather than the prime objective. Good human relations and job satisfaction are deliberately promoted only in so far as they contribute to efficiency. This means that opportunities to play, to do things purely for fun, for the intrinsic satisfaction, tend to be squeezed out of working life. They must be sought outside the workplace, in the after hours. Hence the modern division of life into work and leisure.

In any society where work has been modernized there will be a recognizable domain of leisure. This domain does not extend across all the time that is not accounted for by paid work. There are other things which, like paid jobs, just have to be done. Sleeping, feeding, bathing, household chores and other self-maintenance activities usually fall into this category. Leisure is the time left over. At any rate, it is located in this time. The money that is available to help use this time is the other key resource with which individuals develop leisure interests, engage in leisure activities, and seek fun, diversion, relaxation or whatever experiences they prefer.

The market economy

Everyone who visited the former communist countries became aware that their leisure was different. In the holiday resorts that catered for Western tourists careful efforts were made to provide these visitors with what they were believed to want. There was plenty of pop music, sun bathing or skiing, alcohol and souvenir shops. But when a single state authority operated all the hotels, restaurants and souvenir stores the effect was never quite the same as when a variety of suppliers compete for business. Market economies allow consumers to make more significant choices than were possible under state socialism. In market economies suppliers offer a variety of goods, services and environments. Each business tries to persuade customers that it is offering what they really want. It can be argued that powerful suppliers can sometimes manipulate people's desires, and that markets systematically fail to satisfy the wants of many customers. These arguments are considered fully in later chapters, especially in Chapter 7. The current points are that markets create leisure environments in which individuals feel that they can choose how to use the time and money at their disposal, and this kind of experience is not guaranteed simply by having work-free time and money; it also depends on a surrounding market economy.

Liberal democracy

Multi-party democracies allow voters to choose their governments, but an even more significant feature of these political systems for the character of leisure is that they permit freedom of association. The governments do not try to run everything. Citizens have civil rights. They are able to operate businesses, form churches, political parties and pressure groups, trade unions and professional associations which run their own affairs. And these same civil liberties allow people to organize their own leisure activities in voluntary associations: sports teams and clubs, art societies, gardening clubs, dramatic societies and so on. In any medium-sized town there are hundreds of such associations (see Bishop and Hoggett, 1986). In Toxteth alone, one district in Liverpool, Ruby Dixon (1991) was able to identify 54 voluntary arts organizations. The district appeared to be a hotbed of arts activity, but a systematic investigation would almost certainly have identified just as many sports organizations, and there is no reason to believe that

Toxteth was a particular hotbed.

Voluntary organizations are not profit seeking businesses that are subject to the rules of the market place. Nor are they branches of the state, though they may sometimes receive state subsidies. Voluntary associations are formed and die according to their members' wishes and enthusiasm or apathy. The voluntary sector offers a distinctive kind of leisure experience and adds to the variety of leisure opportunities that are available. Once again, leisure is inevitably different in societies where all organizations are subject to state, or party or church regulation.

The decline of community

The modern organization of work, economic markets and civil rights all conspire to weaken extended families, neighbourhood communities, churches and other belief propagating movements which, in other societies, have prescribed ways of life for everyone. When individuals can obtain their own jobs, locally or elsewhere, and earn their own money, they are able to spend their incomes without necessarily respecting the opinions of other family members, neighbours or church leaders. Voluntary associations and commercial enterprises offer leisure options which may not be part of family, neighbourhood or church traditions. Prior to these modern developments most people had little choice but to share the ways of life of the groups in which they lived, whereas with modern leisure individuals have greater scope to make their own lifestyle choices. Young people can live quite differently from their parents. Members of different households who live side by side can nurture entirely different leisure interests.

The incomplete globalization of modern leisure

The kind of leisure that Western societies make available, and which most of their citizens can take for granted, does not occur naturally, as a straightforward expression of human nature, but has arisen in specific economic, political and social contexts. This kind of leisure has not existed in all societies. What used to be specifically Western leisure is currently becoming more widespread as more countries modernize and as the populations in their modern centres expand. The collapse of communism has also contributed to the globalization of Western leisure. But this kind of leisure is still far from universal. And just as the leisure that we know has not been around from the beginning of history it is unlikely to last until the end of history, whenever that might be.

Western societies are not static; they are inherently dynamic, and possibly never more so than in the late 20th century. They are all affected by the worldwide trends commonly described as globalization, and they are all experiencing waves of technological innovation and economic restructuring which are changing the character of many people's jobs. There are related, though not wholly derived, changes in gender roles and the life course. All these changes make an impact on leisure. They make it necessary to ask repeatedly whether we need to revise our notions about what leisure is. This question can never be regarded as finally answered and it reappears throughout this book. Indeed, nearly every chapter contains challenges to the definition of leisure on which this book is based.

Definitions and how to test them

No one has monopoly rights to decide how a particular word will be used. There are alternative, competing concepts of leisure which are given a full hearing in later chapters and especially in Chapter 6. The definition offered above does not pretend to distil what everyone, or even what all sociologists' mean when they refer to leisure. It is not a dictionary-type definition. It is a stipulative definition which simply explains how the term is being used in the current discourse. It is a sociological definition in that it allows leisure to be defined by its context. Other writers prefer to define leisure from the 'inside'. The sociological concept can be described as residual in that leisure is portrayed as existing in what is left over; the time that remains when paid work and other obligatory activities have been done, and the money that can be spent in that time. Other writers prefer to define leisure by what it is rather than what it is not.

There are three 'litmus tests' that any stipulative definition should survive. First, it should convey the main features of its subject matter. Second, it should be operable – usable in research and in the analysis of the findings. Third, it should be coherently linked to other concepts such as work, market economy, family, civil society, religion and neighbourhood community that intrude into the discourse. The definition of leisure that has been proposed can, in my view, survive all three tests, not unscathed but, it will be argued, in better condition than alternative leisure concepts. Ongoing economic and social changes are constantly presenting the proposed definition with new challenges, but, once again, it will be argued that the challenges can be met without a radical reconceptualization of modern leisure.

Why Study Leisure?

The very idea of taking leisure seriously and studying it systematically still produces sniggers in some quarters. Sceptics wonder why leisure scholars do not switch to something more important and worthwhile. 'Farce of Useless Degrees' was the *Daily Express* front page headline on 21 August, 1996. It was followed up by an editorial deploring

> supposed higher education that offers degrees and diplomas in leisure and recreation studies. . . . Can you imagine for one minute that if university education were paid for directly and in full by parents or students there would be this laughable inflation of subjects?

Twenty years ago this kind of tabloid outcry was quite common. Nowadays it is widely recognized that the jibes are wrongly targeted. It is true that no one needs even to be able to spell leisure in order to enjoy it, but the crucial fact is that leisure is important and needs to be investigated and studied thoroughly for economic, psychological, social and political reasons.

Economic

Leisure today is big business. Tourism alone can claim to be the world's biggest industry. Actually this depends on how size is measured and where the boundaries between industries are drawn. In terms of the numbers who work in it, agriculture beats tourism but not in terms of trade flows. In Britain leisure accounts for between 25% and 38% of all consumer spending, depending on exactly how many objects of spending are treated as leisure (Martin and Mason, 1998). Leisure may be just fun for some but it is a field in which others invest and where millions earn their livings. Leisure is more likely to be a vocational subject for its students than most other social science specialisms. When all the vested interests are taken into account it becomes easier to understand some of the imagery that surrounds leisure activities: why we are told that we will enjoy the experience of a lifetime if we visit, and how our social acceptability will soar if only we drink or wear it, for example.

Leisure is an important source of employment, and it is now one of the few business sectors in which some countries can be confident that employment will continue to grow. These countries include Britain. The economically advanced societies all experienced agricultural revolutions many years ago through which jobs on the land were decimated while, in many cases, agricultural output actually increased. These societies are already well into their second industrial revolutions in which manufacturing is following the path of agriculture: rising output alongside a steep decline in employment. Up to now, in most of these countries, the expansion of service sector employment has fully compensated for job reductions elsewhere, but some service sectors now appear set to follow agriculture and manufacturing. Public sector employment is no longer growing inexorably; voters have become resistant to forever rising taxes. The financial services are introducing computer technology and rationalizing their operations to reduce their labour costs. Leisure remains one of the few economic sectors in which more employment is envisaged almost everywhere. Hence the interest of villages, towns, cities, regions and countries throughout the world in gaining the largest possible share of the leisure market.

The growth of tourism has made it an increasingly important factor in countries' trade balances. No country can afford to ignore the economic impact of inward and outward tourism. Much the same applies to most towns and cities. It is no longer traditional resorts only that want to attract tourists, day trippers and people who are intent on enjoying an evening out. Former industrial, mining and commercial centres are now keen to present their heritage as tourist attractions. They take leisure seriously and are thirsty for information about trends in tourism and leisure more generally, and the features that attract people with time and money to spend. Public leisure services used to be targeted primarily at local populations whereas current local and central governments seem more likely to invest in mega-projects which will attract visitors (see Bianchini and Parkinson, 1993; Street, 1993). In some cases the developers feel it necessary to shield the visitors from the local populations, or train the latter to behave in the manner that the tourists expect, lest the locals blight the attractions. Societies in developing world tourist resorts have become accustomed to this treatment; sections of the developed world population are now tasting the medicine.

Psychological

Leisure has an important economic dimension and it is also important for the well-being of individuals. First, leisure enables people to relax, refresh and literally recreate themselves so that they can return, suitably restored, to other roles in their workplaces and families.

Second, leisure permits people to express desires and drives that would otherwise remain hidden and even suppressed. In everyday language, leisure allows people to 'let off steam' and give vent to their emotions. This is often achieved by playing games. Every game has its own particular rules and there are many different kinds of games, but the rules always serve to separate the activities from the rest of life. This separation is crucial. It means that people can become absorbed in the games that they play. Sports and quizzes can be taken very seriously. The same applies to artistic forms of play. But because it is 'only a game' the outcome does not really matter. People can emerge defeated, or having crushed an opponent, without there being any dramatic implications for anyone's family or working lives. This is why, in leisure, people are able to 'let themselves go', act adventurously, take risks and place themselves in the hands of fate (Vester, 1987).

Third, leisure can be educative. People can develop skills and discover abilities that would otherwise have been untapped. Once again, this is possible because, in leisure, people can experiment and take risks without failure bringing devastating consequences. Through leisure activities, children and adults can develop motor, language and social skills which may then be transported into other areas of their lives. Play methods work in classrooms though this is not to advocate making all education into a game.

Play has performed these functions in all societies throughout the ages. Leisure, as defined above, may be distinctly modern but play is a universal human activity. Some of the classic texts which continue to sparkle decades after they were written dwell on the properties of games (see Callois, 1955) and the vital role of play in civilizations throughout history (Huizinga, 1949). Sociological work on modern leisure has tended to date more rapidly.

The fact that all peoples have found ways and times to relax, recuperate and play, can be cited as justification for defining leisure as a universal feature of people's lives. However, writers who adopt this course invariably recognize that there have been different types of leisure in different types of societies. Rob Lynch and Tony Veal (1996), for example, define modern leisure in much the same terms as this chapter. However, they also identify a type of *eco leisure* which, they claim, occurred in simple societies where leisure was not recognized as a separate sphere of life but where people still engaged in pleasurable and playful activities such as games, music, dance and story telling. Lynch and Veal also identify a type of *pre-modern leisure* which is said to have erupted in irregular bursts, at carnivals for example, when people were able to escape from drudgery to do the same kinds of pleasurable things that featured in eco leisure and which, indeed, feature in leisure today. These authors also identify a form of *classical leisure*, said to have been enjoyed in Greece between 500 BC and 300 BC by a labour-free male elite whose entire lives could be devoted to sport, music and other arts, contemplation and civic participation.

No one disputes that there are similarities between the content of modern leisure and things that people did and experienced in earlier times. Whether this is allowed to lead to an all-embracing definition has to depend on one's judgement of how useful such a leisure concept is likely to prove. Can it survive the 'litmus tests' of conveying the main features, operability and coherence? Beyond this, will a universal definition enable investigators to develop a body of knowledge applicable to 'leisure' in all types of societies? The most reasonable working hypothesis, in my view, has to be that modern leisure is so different from earlier varieties to make it unreasonable to assume that whatever functions modern leisure performs for individuals, groups and society at large will also apply to what Lynch and Veal describe as pre-modern, classical and eco leisure. Likewise, it seems unreasonable to assume that basically the same relationships as those found between uses of modern leisure and other social roles will have been present prior to modernity. Similar relationships are more likely to be found in all modern societies such as in the countries of the European Union, North America and Australasia. This book defines leisure as a specifically modern phenomenon because its analysis and conclusions are intended to apply in all modern societies, but no such claim is made or implied in respect to others. However, this is not to claim that, prior to modernity, people had no experiences similar to those available in modern leisure. Texts that dwell on the psychological significance of leisure have a long and distinguished record of identifying universals, but these are better defined as play and games rather than leisure.

Social

Most leisure has a social dimension and it is therefore socially important. A common feature of leisure activities is their ability to bind people together. Common leisure interests and the related interaction in hobbies, sports and the arts can make groups gel. This can work in families, neighbourhoods, schools and firms, and for cities and countries. The solidarity engendered by shared leisure is likely to be most intense when, as in sport, one group competes against others. Employers, head teachers, urban planners and many other 'social engineers' have tried to benefit from leisure's bonding capacities.

It is important to bear in mind that what can unite can also divide. So home-centred leisure can isolate nuclear families. Leisure based loyalties can set interest groups, schools, cities and countries at 'war' with one another. Provided this is treated as 'just a game' the outcomes need not be distressing for anyone except that the isolation of games from the rest of life is never total. Hence the need to enquire into which groups are strengthened, and which divisions are highlighted and reinforced by a society's leisure activities.

Leisure and the quality of life

Leisure's psychological and social functions make it an important contributor to the quality of people's lives. Here we face a paradox. Most people do not rate their leisure activities as highly important. All the relevant surveys show that most people prioritize their health, jobs and families (see Hall and Perry, 1974; Pronovost, 1988). Popular attitudes to leisure mirror the status of its study in some traditional academic circles. Yet objective measurements invariably find positive relationships between leisure activity and life satisfaction. There is some dispute about the confidence that can be placed in replies to questions such as, 'How satisfied are you with your life?' (see Abrams, 1977). Expressed satisfaction may be a product of low expectations and aspirations. However, for what the measurements are worth (and we have no superior methods of diagnosis), life satisfaction appears to be additive – the product of how satisfied people are with their jobs, places of residence, family lives, health and so on (Andrews and Withey, 1976). And leisure activities are found to be related to general life satisfaction when other contributors are controlled. Indeed, leisure is usually found to be a better predictor of overall life satisfaction than income levels and health.

Leisure's contribution to life satisfaction varies between socio-demographic groups. The relationship is particularly strong among people in later life (see Chapter 5) but it contributes to life satisfaction in all sections of the population (see Kelly *et al.*, 1987; Melendez, 1992). All uses of leisure seem capable of making people feel better provided the leisure is structured, and involves activity and social contact (see Hendry *et al.*, 1993). In other words, going regularly to the cinema with friends does people more good than staying alone at home and switching on a television film for want of anything better to do. Leisure contributes to life satisfaction by providing 'basic' experiences which are beneficial wherever they are obtained – activity, social contact, achieving goals, and being appreciated by others.

Its researchers have not paraded leisure as the all important, exclusive determinant of the quality of life. Leisure researchers are sensitive to the dangers of preoccupation with lifestyle choices diverting attention from other impediments to well-being (see Ingham, 1985). For a start, we know that access to leisure activities depends on people's occupations or lack of any, family situations, gender and age roles (see Chapters 3, 4 and 5). We know that people's jobs and home lives make independent contributions, positive or negative, to the quality of their lives. Leisure research has shown that leisure activities are not satisfactory substitutes for paid employment (see Chapter 3). Jeremy Seabrook (1988) has described vividly how many people's lives have been devastated when economic restructuring in Britain has destroyed the industries and occupations in which all their experience and skills were based. No one claims that leisure can be an entirely adequate replacement.

Robert Lane (1991) has gone so far as to that argue that the 'market experience' must be judged a failure in so far as markets persuade people to accept unsatisfying jobs in exchange for incomes with which they are supposed to be able to lead satisfying lives outside their workplaces. Lane draws attention to the fact that, above the poverty level, there is no clear relationship between income levels and life satisfaction. This is because people's aspirations rise in line with their incomes, and even the higher earners tend to judge the adequacy of their own incomes against those of other people with similar qualifications and jobs to themselves. Lane argues that most employed people obtain most of their 'psychic income' not outside work, but while doing their paid occupations, and that the best way of achieving all-round improvements in the quality of life will be to provide all people with challenging jobs.

It is also the case that people need interests and goals and, in particular, group memberships to celebrate in their leisure which, in itself, appears unable to generate sufficiently powerful and stable bonds to give people a sense of purpose and security. There are occupational communities, neighbourhood communities and religious communities but no leisure communities. Groups with other bases can be strengthened through shared leisure whereas groups that are formed specifically for leisure purposes are far less likely to become multi-functional. People's social and psychological needs must be structured elsewhere otherwise their leisure can be meaningless.

Nicholas Emler and Sarah McNamara (1996) studied the day-to-day levels and patterns of social contact among young people, some of whom were students, some employed, and others unemployed. One of their most interesting findings was that most social contacts occurred through individuals following normal daily routines. Relatively few contacts were made by special arrangement. Equally, few contacts occurred purely by chance. Most arose 'routinely', and because the routines were predictable so were the contacts, though without anyone deliberately arranging to see another particular person. It is through such daily life that most friendships are formed and maintained. Among the young people studied by Emler and McNamara, the students had the most social contacts, followed by the workers. The unemployed were the most isolated. Chapter 3 presents more evidence of leisure itself being unable to compensate when people's lives are stripped of the routines associated with work, whether in education or employment.

However, after accepting all these arguments the fact remains that if people have challenging jobs and otherwise satisfying lives, leisure activities can enhance their well-being further. Leisure is a positive satisfier. Income, in contrast, is more likely to be a source of dissatisfaction. If people feel under-paid this will make them discontented. Pay rises will remove their discontent but will not lift and keep them up the positive rungs of well-being (Phillips, 1967). Leisure activities will not solve all of people's other problems, but once these have been addressed leisure can make their lives even better. People's self-assessed well-being is hardly affected by how they feel about the condition of their countries or governments, but it is highly responsive to their own circumstances: how they feel about their particular jobs, health, homes and family lives, and their leisure. If it is impossible to remove other impediments to well-being – to provide everyone with challenging jobs for example – leisure can ameliorate their situations. If other features of their lives are satisfactory, leisure can improve their well-being even further.

People today have more leisure time, and more money to spend in this time, than in the past. Chapter 2 warns against exaggerating the pace and extent to which leisure has grown, but we have in fact experienced a 'historical inversion' (Dumazedier, 1989). We are now in an era when most people's leisure time exceeds their hours at paid work. Subsequent chapters explain how recent economic and social changes have destabilized many people's occupational careers, neighbourhood and family lives. These have become less reliable for many people. Such trends underline the importance of recognizing, and using, leisure's ability to contribute to personal and social well-being.

There is a similar relationship between leisure and physical health. Access to medical care is obviously important for the maintenance of physical well-being. Working and housing conditions are known to have powerful effects on 'health status': people's long-term health condition as opposed to the week-to-week variations that occur in everyone's state of health. However, the foods that we eat, our use of alcohol, whether we smoke, and amount of exercise make an independent difference to our physical health. Lifestyles cannot override biology. Nor can they override the injuries inflicted by unhealthy living and working conditions (see Blaxter, 1990). The effectiveness of health promoting pastimes can be exaggerated. For example, if people wish to use exercise to improve their health they need to exercise strenuously at least three times per week and maintain this routine for life if they wish to retain life-long benefits. Less frequent exercise usually improves individuals' fitness to undertake the relevant activities but without achieving health promoting changes in their bodies' physiological functioning. Healthy diets, not smoking, and not drinking heavily usually achieve sharper improvements in health status, though without eliminating the inequalities associated with other health determinants (see Roberts and Brodie, 1992).

Leisure is not a cure-all but once people have access to the best medical attention, and when they have achieved benign working and living conditions, their lifestyles are likely to become major discriminators of health status. People in these conditions seem aware of this. The pursuit of health has become a prominent lifestyle goal among sections of the upper middle classes (see Savage *et al.*, 1992). They are less likely to be victims of a postmodern insecurity (see Chapter 8) than people who simply realize that how they spend their leisure can affect their health, and that lifestyle factors are health determinants over which they can exercise direct and immediate control.

Political

All governments become interested in leisure if only because of the functions described above. Nowadays no government can afford to be indifferent to the economic significance of leisure. Most national governments have tourism policies and programmes usually designed to promote the import of tourists. Local government bodies are most likely to have similar concerns and policies.

Governments have become equally alert to the value of leisure in social bonding. Specifically, governments have sought to foster national solidarity and with this, they have usually hoped, support for their regimes. Pride in national heritage and culture can help to strengthen if not create a sense of national identity. International sporting success has been pursued by governments as a means of raising their countries' profiles and boosting national esteem. However, such measures may be insufficient to save otherwise unpopular regimes. Some communist states were spectacularly successful in international (especially Olympic) sports but most of the governments and their political systems have now been replaced (see Riordan, 1995). There are limits to what leisure can do for governments, and (see below) what governments can do with leisure.

Some governments have also been alert to the functions of leisure for individuals and have sought to provide all sections of their populations with opportunities to participate in 'desirable' leisure activities. In the 1960s and 1970s there were some signs of leisure gaining recognition as a branch of the welfare state and of all citizens becoming entitled, as of right, to take part in sport, appreciate the arts, visit the countryside, and even go on holidays away from home (Bramham and Henry, 1985; Henry, 1993). Today it seems doubtful whether any governments, in the foreseeable future, will be able to raise the funds to bear the costs of such aspirations.

It can also be queried whether governments are able to decide what leisure is desirable for all their citizens. Experience under communism has shown that governments cannot cater for leisure with quite the same effects as voluntary associations and commercial markets. Public leisure provisions are always liable to reflect what politicians, or the classes of people who the politicians represent, consider good for other people. Hence the (valid) criticism that the UK National Lottery is a device for persuading poorer people to pay for richer people's pleasures and good causes. State support for participant sport, which has usually been considered especially beneficial for young males, has been explained as a means of nurturing compliant workers and soldiers (see Clarke and Critcher, 1985; Hargreaves, 1986; van Moorst, 1982). In other words, state sponsored leisure has been interpreted as social control. Are there enough votes in such state interventions for elected politicians to remain enthusiastic? Another limiting consideration is that in societies where wealth and income, together with standards and styles of living, play key roles in socio-economic regulation, it is doubtful whether recreational welfare rights could ever be pitched at levels that most people would consider satisfactory without incurring huge costs in socio-economic efficiency.

Governments inevitably become involved in leisure irrespective of whether they wish to encourage particular uses of free time if only because they are the ultimate custodians of social order and, 'Leisure is part of the struggle for the control of space and time in which social groups are continuously engaged . . .' (Wilson, 1988, p. 12). Leisure time and money have not been delivered automatically to the people by economic growth. They have usually had to be won in labour markets, by bargaining with employers, or by persuading governments to regulate hours of work and rat es of pay (see Cunningham, 1980; Rosenweig, 1983). Governments must decide which demands to endorse, and they must define the limits of what is tolerable, and may need, if only as a last resort, to arbitrate between competing claims on resources – for the use of water space by anglers and skiers for example, or to protect coastal or rural areas from industrial encroachment, or to protect sensitive citizens from spectacles that they find offensive. Leisure is unavoidably political. Various groups constantly stake claims on physical and socio-cultural space, sometimes within the existing politico-legal framework, and at other times by resisting its demands and seeking changes. One group's innocent pleasure may offend another. In Britain roughly a third of young people are regular users of illegal drugs at some stage in their lives, and another third are occasional users (Measham *et al.*, 1994). Pornography is another huge, and highly controversial leisure industry (see Box 1.1).

Governments may also intervene in leisure because there are specific pastimes that influential politicians are personally keen to promote, or because politicians have vested interests in servicing the leisure interests of particular sections of the public thereby earning their gratitude. Some researchers have sought the rationale behind government involvement in leisure, but there proves to be no single explanation (see Coalter *et al.*, 1988). Public provision will do whatever politicians sanction, subject to the normal checks and balances of political processes.

If leisure becomes more important to members of the public, not just objectively but in their estimations, its political importance is sure to grow. This is not to envisage parties forming around leisure interests though more people may become involved in pressure groups, so-called new social movements, to protect the countryside, pets, playing fields or whatever they

Box 1.1. Pornography.

This is a very large leisure industry. The main product is now the sex video. In the USA 150 new titles are produced every week. USA pornography has a larger turnover than:

- mainstream cinema,
- theatre,
- rock and country music.

Source: Sharkey (1997).

value. Leisure will probably remain non-party political if only because leisure interests tend to cross-cut the other social alignments that have determined party support up to now. However, all parties could encounter growing pressure to cater for leisure interests, and realize that there are marginal but possibly crucial votes to be won or lost depending on their stances on leisure issues – on opportunities to play sport, enjoy the classical arts, roam the countryside or beaches, or to have favourite programmes on the television channels that most people receive. Sulkanen (1997) argues that as 'consumption' becomes more important to people, there is more and more pressure on governments to deregulate activities that people find pleasurable (drinking alcohol and gambling for example). Simultaneously, we demand that governments protect us from other people's habits which we find offensive (tobacco smoke for instance). We also demand that governments supply us with reliable information (about food products among other things) so that we can pursue our own pleasures safely.

Up to now most public leisure spending in Britain has been by local, not central government. Leisure has in fact been the largest element in many a local authority's budget. Few members of the public may be aware of this because of their limited interest and involvement in local politics. In UK local elections people tend to vote, if they bother to do so, according to their national party sympathies. However, since 1992 most central government leisure spending has been channelled through a single department, originally the Department of National Heritage, and since 1997 the Department for Culture, Media and Sport, which is also responsible for the National Lottery. This is big money and how it is spent easily becomes politically controversial – how the benefits are distributed between regions, between projects which are majority and minority, and mainly middle class as opposed to working class interests, for example. Electoral pressures or their own financial problems have already persuaded politicians that some Lottery funds should go to health and education rather than leisure facilities which, as explained above, are not generally regarded as crucial, whatever the objective evidence says, to people's quality of life.

Mulgan and Wilkinson (1995) have forecast that time will become a major political issue of the 21st century. They have in mind people demanding parental leave, sabbaticals for re-education and training, and ceilings on the hours that they can be expected to work in order to be considered serious candidates for promotion. During the 20th century governments have come to be regarded as responsible for production issues – the extent to which economic output and standards of living grow. Governments are held equally responsible for the overall distribution of society's resources and these include time as well as money.

Where the main political alignments are drawn in the future will depend, as in the past, on the groups with which voters identify most closely and from which they derive their identities and conceptions of their main interests. Chapter 8 considers whether leisure is likely to take on this role, but even if this happened it is doubtful whether political parties would be joined and votes cast primarily on the basis of leisure policies. This is because while governments can make a crucial difference, if they choose to do so, to most people's housing and educational opportunities, medical services, employment prospects and pensions, they have far less leverage over how people spend their leisure. Government policies can make some difference to how leisure is spent but in Western countries governments are anything but all powerful. It was different under communism when the state and party were the only major providers (see Riordan, 1980, 1982). In Western countries people can ignore public provisions and use the market or voluntary sectors. Even under communism people had the option of organizing their leisure privately, in their homes with friends and family members. If people do not like a government's leisure policies they have the soft option of simply ignoring public provisions instead of seeking to change the government or its policies. For most people it would be more difficult to make alternative arrangements for their families' education or health care. How people spend their leisure ultimately depends on what they want to do rather than what a government or any other sector is providing. This is why the most powerful explanations of leisure behaviour start with types of people rather than types of provisions. In the following chapters it will become clear that in Western countries no single provider, or even a set of providers, has the power to stamp a decisive pattern on the people's leisure.

Despite their inevitably limited influence, politics and governments can never be kept out of leisure entirely. And the more leisure people have, the greater the economic, psychological, social and political importance of leisure becomes. These are the justifications for studying leisure. It is an important part of modern societies. Leisure needs to be understood as part of a comprehensive appreciation of how modern societies are structured and how they function. The information that leisure research can produce may prove useful to commercial businesses, leisure interest groups, individuals and governments, but it is pointless to raise unrealistic expectations. The applications always depend on the findings. The basic reason for studying leisure is simply that we 'want to know'. Uses of research findings are never predictable. They always depend on potential users discovering something in the evidence that is relevant to their own purposes. The basic goals of research, and this book, are simply to describe, analyse and explain.

Basic Information about Leisure

The study of leisure is not an infant discipline. It may not have attained the size or stature of the study of work, politics or education but it has almost as long a history. And nowadays leisure researchers face not a shortage but a wealth of basic information about the population's leisure activities, time and spending. These data mountains grow so fast that the material is never analysed exhaustively. There may be a shortage of small and medium-scale studies that explore the fine detail of the leisure of specific social groups, and carefully designed enquiries into specific leisure issues, but there is a wealth of information about the distribution of leisure time, leisure activities and spending throughout the entire population in the UK and most other modern societies. Most of this information has not been gathered to satisfy someone's disinterested thirst for knowledge. It has been collected routinely, usually at the behest of public and commercial leisure providers who want to establish the actual and potential sizes of their markets and audiences.

Participation surveys

These provide one kind of information about the population's leisure. Participation surveys usually present their samples with checklists of leisure activities and ask which have been done within a given time period, and how often. The main British survey that has collected this type of information is the General Household Survey. This is conducted annually and covers a nationally representative sample of households. It is a general, multi-purpose survey, and carries any questions that government departments (and sometimes outside bodies) wish to have included, and are willing to pay for. Since 1973, and every subsequent third or fourth year, the General Household Survey has included a battery of questions about leisure activities. It records, for example, the proportions of the population that make countryside trips, visit museums and play various sports. Participation rates for different sections of the population can be calculated and compared, and since these leisure questions have been asked periodically since 1973 trends over time can be established.

Presentations of this type of information are normally, and justifiably, accompanied by hazard warnings. The evidence is not quite as hard as it can appear in tables of figures. The percentages shown to be taking part depend on the thresholds for being considered a sport participant for example; at least once a month, less or more frequently. The specific activities that are included under certain broad headings also repay scrutiny. For example, the General Household Survey's list of sports contains non-strenuous activities such as darts and snooker, and all types of physical recreation are included – recreational swimming and walking for example – not just those cases where there are sporting contests. Participation surveys are more discriminating for some activities than with others. The measurements work quite well with structured activities such as sport which may be played daily, weekly or more occasionally, but the findings are less useful in respect of activities such as watching television which most people do daily, and with unstructured pastimes such as listening to radio music in which people can have difficulty in recalling exactly how often they have participated.

Time budget studies

These have been conducted rather less frequently, and less regularly than national participation surveys. Time budgets are expensive to administer and analyse, especially with large samples. They require subjects to keep diaries for a week or a fortnight, recording their main activities in each 15 or 30 minute period, and usually seek additional information such as where the subjects were and who they were with. These enquiries have often been sponsored by the broadcasting media which have vested interests in discovering not only how many and which people watch television or listen to the radio at given times, but also who is at home or driving their cars and therefore in the potential audiences.

Time budget evidence permits calculations of the total amount of leisure time that is available to the population as a whole, and to various sub-groups. The proportions and amounts of leisure time devoted to specific activities can also be calculated. This kind of evidence discriminates levels of television viewing and radio listening, which many people do daily, more effectively than participation surveys, but time budgets have their own limitations. Because it is unrealistic to expect people to maintain diaries for longer than two weeks, the method is likely to miss more occasional activities such as holidays. The method is also liable to miss activities that are not precisely time bounded and which are often done alongside something else. Talking with other family members which may happen while preparing or eating meals or watching television is one example.

Economic statistics

There is a wealth of information about the public's spending patterns, the most detailed in the UK being from the Family Expenditure Survey. Like the General Household Survey, this is conducted annually with a large, nationally representative sample. Like time budgets, it requires the subjects to keep records, of all their spending in this case, and to give details about their incomes. From this information it is possible to calculate the total amount of leisure spending, and the amounts devoted to specific types of leisure. As with participation surveys and time budgets, these calculations can be made for the population as a whole, then sub-divided in numerous ways.

This data also has inherent limitations. All survey evidence is subject to errors arising from respondents' imperfect memories, or their failure to disclose some of the things that they have done. We know that people tend to under-estimate their spending on items often considered wasteful such as alcohol and tobacco, and grossly exaggerate their participation in 'worthwhile' activities such as sport (see Cale and Almond, 1992; Chase and Godbey, 1983). Economic statistics have additional blind spots. They do not pick up free uses of leisure such as relaxing. Also, there are many types of spending within which it is difficult, in practice impossible, to separate the leisure element. Spending on clothing, transport and telephones are examples.

Major uses of leisure

All the above types of information have characteristic limitations, but by combining the different kinds of evidence it is possible to identify the public's main leisure activities: things that most people do, and which account for considerable proportions of their leisure time and/or spending. The big three that are identifiable in this way are the media, eating and drinking, and holiday making. Eating and drinking head the spending league, tourism is in second place, and home entertainment (mostly TV-related) is third (see Table 1.1).

In terms of time use, home entertainment is first placed. In the 1990s adults in Britain have been watching television for well over 20 hours per week on average. If time spent attending to all the mass media is aggregated, it amounts to approximately half of all leisure time. The media play a major role in most people's leisure and, since the 1950s, among the media, television has been by far the most important, certainly in terms of time accounted for. The media are also major objects of leisure spending. People

Table 1.1. Consumer spending on free time, UK 1996.

	£bn	% Change, 1971–1996
Eating and drinking	53.8	46
Holidays and tourism	23.9	213
Home entertainment	13.2	676
Home and garden	9.5	166
Hobbies and pastimes	6.5	99
Gambling	5.6	33
Reading	5.5	–11
Sport	4.8	193
Local entertainment	3.3	53
All free time	126.0	94

Source: Martin and Mason (1998).

may regard switching on the television or the radio as free entertainment but in fact they spend substantial sums, relative to other leisure items, on renting or purchasing the sets, on cable and satellite reception, and video recordings. The home is most people's main leisure centre. The family is their main leisure group. And when at home, with their families, people are more likely to watch television, or attend to other media, than engage in any other leisure activities.

When people go out, the thing that they are most likely to do, and spend money on, is to consume food and drink (especially the alcoholic varieties). Participation in these activities, and the sums spent, vastly exceed the figures for cinema admissions, and sport spectating and playing. When people go out in the evenings or at weekends their most likely activities are eating and drinking. Sometimes these are the central events in the outings but people are also likely to purchase and consume food and alcohol when they go out for other leisure purposes – for day trips to the countryside or to concerts for example. The result is that out-of-home food and drink are leading leisure items.

Tourism is the third member in leisure's leading trio. Approximately two-thirds of the population take at least one holiday away from home each year. Participation is not as high as for television viewing and alcohol consumption, and tourism is not outstanding in terms of time accounted for since most people 'go away' only once or twice a year. In terms of expenditure, however, tourism is definitely among the population's leading uses of leisure. Over a 12 month period, the typical household spends more on its holidays than on all the audio-visual media. Over time, spending on holidays has been rising more steeply than most other kinds of leisure spending, though home entertainment heads this league table (see Table 1.1). People have been taking more holidays, and travelling further. The extent to which people are willing to save up for the greater part of the year in order to enjoy a week or two 'away' suggests that these occasions must be highlights to which individuals attach considerable importance.

Is the Basic Information a Suitable Foundation?

Solid achievements

The data from the major time budget, participation and expenditure surveys have been gathered without any charge to academic researchers. We have not needed to make research applications or raid our own budgets unless we have wanted to do further work on the data sets. The basic findings, the national picture, have been available gratis. This picture can be used as a basis for further research, but should the data be given so pivotal a position?

The data is 'hard', and appears factual and neutral in the columns of statistics. The foundation has been resilient in that the main contours of leisure participation, time use and spending have not changed dramatically in recent times. For example, attending to the media has accounted for a major block of leisure time since the 1930s. Before the age of television much more time was spent listening to the radio and visiting cinemas. Moreover, the main contours have proved common in all the modern societies where comparable data have been gathered (see Cushman *et al.*, 1996; Stockdale *et al.*, 1996). Everywhere television is a major time filler and, as standards of living have risen, tourism has always become a major area of leisure spending. Another common feature has been the long-term growth in leisure time, spending and activity.

Some major contrasts between socio-demographic groups have also persisted over time, and been found in all modern countries. For example, men always play more sport than women and drink more alcohol; the higher socio-economic strata always do more of most things, the main exception being television viewing; and young people are the most active age group in most forms of out-of-home recreation. Needless to say, there are endless subtle variations between countries, and some sex, age and social class differences have shifted significantly over time, but not sufficiently to shatter the main macro-patterns. Leisure research has laid or, at any rate, appears to have laid a solid base on which further investigations can build by enquiring, for example, exactly whose leisure has grown; why the patterns and variations between socio-demographic groups are as they are, adding fine detail to the macro-picture; and explaining how and why the leisure of specific groups of men and women, young and old, and workers in specific occupations, departs from general and normal patterns.

Up to now the study of leisure has been a success story. It is certainly not a failed or floundering project. It has been a success in terms of the growth of research, publications and students, and in producing robust findings on how modern societies produce their type of leisure, and how its distribution and uses vary between socio-demographic groups. Some of the problems that confront leisure researchers today are products of this success. It has become difficult to keep pace with the flood of information. Also, leisure research has been so successful in answering its basic questions that it has become difficult to produce original findings on these issues.

Leisure research has succeeded so well that it has nurtured substantial sub-disciplines: the study of sport, the media and tourism for example. There are some advantages in sub-dividing leisure. The phenomena being dealt with become less diverse, and all branches of the leisure industry are interested in learning more about their particular markets. However, it is through studying leisure as a whole that the most powerful explanations are developed. This is because society is not divided into sport players, television viewers, tourists and so on. It is the same people who do all these things. And the inability of any suppliers to impose their preferred patterns on leisure behaviour means that the most powerful explanations of what people do are in terms of types of people rather than types of provision. Moreover, these powerful explanations are common across all areas of leisure. The study of leisure has been successful in establishing the basic social geography of what people do, but this does not mean that the project should be considered complete. Rather, the foundations have been laid from which to address 21st century issues.

Questioning the foundations

Chapters 2 to 5 of this book build squarely upon the foundations. The evidence from time budgets, participation and spending surveys supplies the core. The historical growth of leisure is examined in Chapter 2, then subsequent chapters examine variations according to people's types of work, or lack of any, between males and females, and between age groups. These chapters explain how successive waves of research have deepened our understanding of how people's leisure is related to their wider circumstances. For example, researchers have become more sensitive to the numerous ways in which gender roles have leisure effects, how gender roles themselves differ among both men and among women, and how leisure is not just affected by, but can have consequences for, the social construction of masculinity and femininity.

However, gender roles, work roles and the social significance of age have not remained constant, so researchers' questions have been re-appraised, and in some cases they currently need further re-appraisal to cope with ongoing economic and social trends. Social change is a theme of this book. We have become a post-industrial society, at least in the sense that less than one-sixth of employees in Britain today still earn their livings in manual jobs in which they manufacture things or extract materials from the earth. And people's experiences in jobs of all types have changed as the economy has become post-Fordist (see Chapter 3). Alongside these trends there have been changes in gender and age roles. Simultaneously, the overall volumes of work free-time, economic output, and therefore consumption, leisure activity and spending, have risen. This book is about the implications of these trends. Some of the trends have been long term. However, in the case of the growth of leisure we have recently been reminded that even long-term trends need not continue indefinitely. It has been discovered that the groups for whom leisure is still growing, and the form that this growth is taking, are not quite the same as earlier in the 20th century.

Modern leisure is a product of modern societies, and the broader organization of social and economic life, and trends therein, have always posed the big questions for leisure research. As societies modernized, these questions arose from the wider implications of, for example, the kinds of occupations, gender divisions and age roles that were created. Leisure research's key issues have always been posed by broader socio-economic trends, and this is the basic reason why the study of leisure benefits when it is part of broader debates about the organization of economic life, class divisions, gender inequalities, and so forth. Trying to go it alone as autonomous leisure studies, developing a closed body of theory about leisure behaviour, motivations and satisfactions, can be a death sentence. Leisure research is healthiest when it is an outward looking part of larger projects.

The following chapters advocate building on the foundations laid by leisure research up to now, but not by continuing to address exactly the same old issues. The chapters also acknowledge the need to reappraise the foundations constantly. For example, they show that examinations of present-day leisure constantly encounter problems many of which, it can be argued, stem from the definition of modern leisure from which this chapter began. The definition proves difficult to apply among some sections of the population that have grown in size in recent decades; the unemployed and the retired for example, as well as among women regardless of their employment status. The findings from some detailed studies have fuelled critics' arguments that the portraits of leisure obtained from large-scale surveys of time use, participation and spending are largely artefacts of the methods of measurement. Many of the critics would prefer to begin from the inside with actual leisure behaviour, its immediate contexts, and the actors' motivations, experiences, gratifications and frustrations.

How people use television and alcohol

Take television viewing for example: there have now been a number of studies not of how much people view but what they actually do when they watch. An excellent example is Paivi Pontinen's (1996) interviews, which lasted an average of 2.5 hours, with thirty 17–19 year olds in Finland, and explored their uses of all the audio-visual media. Pontinen found, as expected, that the young people were heavy consumers of media output. When at home they would normally have something switched on. They experienced silence as abnormal. But they would rarely do nothing except watch television or listen to the radio unless they had nothing better to do. The most common exceptions were when the young people watched televised films. Otherwise listening and viewing were not among their leisure priorities: friends had pride of place. On most occasions when they listened or viewed the young people would also be doing something else: eating, drinking, talking, reading or writing essays for example. They normally switched on without knowing what was on, then channel surfed until they found something worth watching or listening to which would usually be partly over. It seems not only bland but a gross distortion to describe all the occasions when these young people had something 'on' as having been spent watching television or listening to the radio, tapes, records or compact discs.

Likewise when people go out for a drink the most common reported attraction is not the alcohol but the company of friends (Wilders, 1975). How accurate is it to describe 'drink' as the object of this leisure spending and time use? In a similar vein it can be argued that most people have leisure interests and projects that overlap and combine the categories normally used to analyse survey data. For example, a person who is keen on sport may follow it on television, radio and in newspapers, talk about it with friends, play and attend fixtures as a spectator, and even go on sports oriented holidays. It certainly distorts reality to count as 'sport' only the occasions when such a person actually plays or watches as part of a live audience, and to describe the other parts of his or her leisure as devoted to the media and tourism. There are many different types of media consumption and holiday making, and it can be argued that these categories need to be disaggregated, and sometimes combined with ostensibly different leisure activities, before trying to explain anything.

New agendas

Chapters 6 to 8 introduce some recently developed agendas for leisure research which give more prominence to leisure experiences and processes themselves. It can be argued that these perspectives offer a better appreciation of the role of leisure in present-day societies, which this book describes as post-industrial but which other writers describe as actually or incipiently postmodern, than measurements of people's quantities of leisure time and spending. Chapter 6 considers systematically the advantages and disadvantages of abandoning residual, social structural, 'modernist' concepts of leisure in favour of social psychological definitions as a kind of experience. It concludes, however, that although there have always been problems in operating with the residual concept, in modern societies this has always been, and remains, more usable and productive than any of the alternatives. It will be shown that experiential concepts have weaknesses of their own. They are difficult to operate in field research, so the evidence with which the concepts are associated tends to be anecdotal and the arguments that arise tend to be speculative. This can lead to exaggerated claims about leisure's current role and importance.

The economic and social changes that are in process, and which feature throughout this book, raise questions about whether present-day societies are still modern. Chapters 6, 7 and 8 deal with claims that we are currently entering a postmodern age, and/or that modernist modes of thought, of which the conventional study of leisure might be used as a prime example, are now outdated. Some writers believe that the concept of leisure has become dispensible and would absorb the subject into cultural studies or analyses of consumption. Postmodernization is usually seen as arising, at least in part, from a weakening of the structures and divisions which formerly defined working life, gender and age roles, and leisure time and activities. In some visions of the postmodern age leisure disappears in a surrounding flux. In other versions it acquires greater importance as, through consumption and other uses of leisure, people fashion lifestyles on which they base their self-concepts and principal social identities. These are challenging claims but, as Dominic Strinati has observed, 'There has, in fact, been a tendency to assume that postmodernism has become widespread in modern societies. However, less attention has been devoted to demonstrating that this is the case' (Strinati, 1995, p. 223). This book will show that the study of leisure can supply, and in fact has produced the evidence to test these claims. This evidence is considered mainly in Chapters 7 and 8 in the course of which the more extreme claims about leisure's disappearance and enlarged role are rejected.

Post-industrial but still modern

The following chapters accept and indeed are based on an acceptance that countries such as Britain have become post-industrial and that economic life has become post-Fordist; but they reject the postmodernity hypothesis. The trends that postmodernist writers highlight are real. Older structures and divisions have become more fragile. Consumption has expanded and consumer culture has become stronger and more widespread. But these are all just trends, not absolute states, in societies which remain generically modern. It will be shown that the best way to investigate the implications of these trends, and to interpret the evidence, is to build on, not jettison the past achievements of leisure research. Leisure remains important for the economic, psychological, social and political reasons that have applied throughout modernity, and which were outlined earlier. Leisure is becoming more important because it is growing in the ways conventionally described – in terms of the time that is available and the money that people can use in this time, and therefore the number of activities in which they can participate. These are not just sufficient, but sound reasons for studying leisure.

Chapter 2

The Growth of Leisure

The Rise and Rise of Modern Leisure

It has been customary, rightly so, for books on leisure to open by stressing the subject's importance: such regard for leisure still cannot be taken for granted. Subsequently the books normally and correctly explain how over time leisure has grown ever more important; that it has become a larger, more significant, more central element in people's day-to-day lives and in their overall quality of life. This book is true to type in all these respects. The growth of leisure is endemic in modern societies. Having created leisure by making work into a part of life, the economies generate more and more. This long-term trend is ongoing and no end is in sight.

All the quantitative indicators reveal long-term growth, in leisure time, leisure spending and participation rates in leisure activities. It is relatively easy to demonstrate leisure's quantitative growth. It is not quite as straightforward, but no less plausible, to postulate related qualitative trends: towards people's well-being – physical, mental and social – becoming increasingly dependent on their uses of leisure; towards leisure interests and activities becoming more central and pivotal in people's overall ways of life; towards people becoming more concerned about, and able to protect their leisure by tailoring work and family commitments accordingly; and towards leisure becoming a base for people's self-concepts and social identities – how they regard themselves and how they are regarded by others. All this easily becomes an agreed credo among people with vested interests in leisure because they cater for other people's or study and write about it. In fact, as the previous chapter suggested, some of the propositions are extremely controversial, less firmly established than they can be made to sound. This book is sceptical towards some of these claims and discriminating about the senses in which leisure is becoming qualitatively more important.

The conventional (and correct) story of leisure's long-term growth invariably starts with the development of modern industries. In Britain this occurred between 1750 and 1850 when the country was transformed from a mainly rural into a predominantly urban society, and when manufacturing replaced agriculture as the main economic activity. The far reaching changes that took place in this period are usually summarized as 'the industrial revolution'.

Rolling back hours of work

All students of modern history learn about the long hours and hard labour that were imposed upon their workforces by the early industrialists (Cunningham, 1980; Malcolmson, 1973). During the first half of the 19th century a normal working day in the factories could last for 12 hours or more. In winter time people commenced work before dawn and continued after sunset. This was before the age of electric lighting. At that time manual labour was often extremely heavy. Many tasks had still to be mechanized let alone automated. Railway cuttings were excavated using human energy. During the industrial revolution the normal working week was extended to 6 days and Sunday remained the only generally recognized day of rest. Employers waged a long and ultimately unsuccessful struggle to discipline labour to accept such regimes.

The problem was not just that employers were seeking to impose long hours of work but also a new rhythm of working life. Previously most people's work had been governed by 'natural rhythms' (see Thompson, 1967). Work was done when it could be done or had to be done. In agriculture there were periods such as harvests when there was much for everyone to do but for most of the year the pace of life was more relaxed. The traditional English weekend had lasted 2 days, usually Sunday and Monday, and the calendar was liberally sprinkled with saints' and feast days. People had participated in rural pastimes and sports, found amusement at fairs, and alcohol had flowed freely on all occasions of celebration. This was the background to the 19th century campaigns for temperance, to encourage the working class to abandon traditional amusements and adopt sober ways of life (see Malcolmson, 1973). Employers struggled to discipline workforces that continued to recognize 'St Monday' and annual 'wakes', and who spent their free time in debauchery rather than physical and mental recovery. One advantage in employing child labour, according to some employers, was that the very young were easily moulded to industry's requirements. Adults who migrated from the countryside bringing their blood sports and drinking with them were less pliable. If children were too young to be employed it was considered desirable that they should be schooled in the habits of discipline, punctuality and industriousness. These habits were in fact taught in the elementary schools, most of which were run by the churches, otherwise by private individuals, then, after 1870, by the new local School Boards. The churches and the schools partnered employers and the temperance movements in propagating a work ethic consistent with industrial employment.

Prior to 1850 paid holidays were simply not part of industrial life. If workers continued to recognize traditional holiday weeks their time off was unpaid. Most traditional saints' and feast days had no place in the new industrial calendar. Between 1761 and 1834 bank holidays were reduced from 47 to just 4 (Myerscough, 1974). Men, women and children were subjected to industrial hours and discipline in the mines and factories. There was no compulsory education nor old age pensions. People would work for as long as they were able. Thereafter they became dependent on their families, charity or, as a last resort, the workhouse. Poor relief operated on the principle of 'less eligibility'. In order to motivate and discipline labour it was considered essential that the conditions of those who resorted to charity or public assistance should be distinctly less attractive than those of the lowest paid worker.

It would be misleading to create an impression that everyone worked 12 hours per day, 6 days a week, 52 weeks a year, from childhood to infirmity. For a start, employers could not guarantee regular work. Second, the workforces often preserved their traditional leisure occasions. This was in addition to individual absences. Third, many young children were kept out of the factories and mines, and were sent to school or remained at home assisting with the care of younger siblings. For many women, child-bearing and child-rearing were necessarily full-time occupations. Subsequently employment has become more regular and workforces have become better disciplined, and these trends will have offset the impact of reductions in the standard working day, week and year, which has been one of three major trends responsible for the long-term growth of leisure.

After the mid-19th century standard hours of work began to be rolled back and the long upward march of modern leisure commenced. There were several forces instigating this trend. First, legislators began to restrict hours of work, initially for vulnerable groups such as women and children, and in industries which were considered especially hazardous such as coal mining. Legal ceilings were introduced on the length of a working day and employers were required to give their staff breaks during otherwise long unbroken working stretches. National opinion was shocked by the reports of the Factory Inspectors on conditions of work in the mines and mills of industrial England. Child labour became an issue, a crusade, that lasted into the 20th century. Women and children had worked in rural Britain but this had usually been in the family or in family-like environments in the cases of domestic servants and apprentices. What outraged 19th century opinion was not so much that women and children were working, but the newconditions including the long hours of work that were being imposed by employers, most of whom could not be described as paternal.

Trade unions were the second force behind the reduction in basic hours of work. Craft workers began to organize effectively in 'new model unions' from the 1850s onwards, and trade unionism spread to non-skilled labour in the 1880s. As soon as labour began to organize and confront employers with collective demands these invariably included higher pay and shorter hours. It has remained so to the present day. Reductions in working time have not normally been bestowed on grateful workforces by benevolent employers. Such improvements in terms and conditions of work have usually been outcomes of workforce organization and struggle. Working hours have not been reduced gradually and smoothly from year to year. There have been major lurches, separated by years of stability. The downward lurches have usually coincided with upturns in the business cycle, when employers have been able to afford concessions. However, the employers have invariably needed exceptional pressure from trade unions, often generated by fear of unemployment, the experience of recent recessions having remained at the forefront of negotiators' minds (see Bienefeld, 1972).

It should be noted that in bargaining for reduced hours of work, labour was implicitly accepting the 'payment for time' package. The industrial workforce came to accept industrial work discipline – the need to work reliably – in exchange for fair rewards. Labour learnt that time was money and tried to use the formula for its own advantage. Everything could be given a price. Shift systems and overtime proved acceptable if and when the price was right. By the time that trade unions were successfully negotiating for shorter hours, their employers' battles to industrialize their workforces had been won.

The third force behind the reduction of standard working hours was that from the mid-19th century onwards some employers began to adopt 'enlightened', 'progressive' approaches. Many realized that tired workers were neither efficient nor reliable and that the loss of working time could be at least partly compensated by increased productivity and fewer mistakes during the remaining hours of work. This became a more weighty consideration as industry became more capital intensive and mistakes became more expensive and dangerous. When 'Taylorism' spread during the 20th century – so-called scientific management with its stop watches and work study methods – employers became aware, because their own methods were producing the evidence, that long hours could be counterproductive. They discovered that work regimes could be intensified and quality standards maintained only by giving their workforces regular breaks, and limiting their total hours of employment.

Moreover, employers began to be aware that using leisure to motivate their workforces could be more effective than trying to suppress their pastimes. Employees, it was realized, could be persuaded to attend regularly and punctually, and work hard, in exchange for the promise of leisure. Needless to say, the employers wanted to ensure that their workers' free time was used 'sensibly'. Instead of just opposing the undesirable, 19th century employers joined the churches and other social campaigners in urging 'rational recreation' (see Bailey, 1978). Workers were encouraged to adopt improving, edifying pastimes in preference to their bloodsports and drink. Edifying pastimes included using the libraries that were being opened, attending the evening classes that were becoming available, going to the churches that were being built, and playing the modern sports that were initially invented in the public schools and universities then taken to the masses. Some employers developed work-based recreation. A few built their own model industrial villages. These campaigns were not 100% successful but they were a factor in the development of 'respectable' lifestyles among sections of the working classes, generally the more skilled and better paid, and those whose jobs were the most secure (see Bailey, 1978). The middle classes more generally sought to use local government to civilize the industrial cities by laying out parks and playing fields, opening art galleries and museums, and building concert halls. There were hopes that these facilities would civilize the labouring classes but by the time of the First World War the middle classes were in full flight to the suburbs having largely abandoned their earlier aspirations (see Meller, 1976). Suburban living became an option as suburban railways were opened, and tram and bus services were introduced.

All the edges of working time have been trimmed. The standard working day was cut back to no more than 10 hours in nearly all industries during the second half of the 19th century. Subsequently it has been reduced to its current norm of 7 or 8 hours in a full working day. The contraction of the normal working day may now have ended. At some point the time and money costs of travelling to work must cease to be recompensed from the worker's viewpoint. Employers too are bound to find at some point that their workforces can remain fully alert for a full working day and there are no benefits to offset further reductions in working time.

The standard working week also began to shrink in the second half of the 19th century. An extra half day of leisure was won. Saturday afternoons became leisure occasions. A reason why some employers preferred to release part of Saturday rather than restore 'St Monday' was that if workers became intoxicated on the Saturday they could recover on the Sunday. The 5½-day week remained the norm throughout industry until after the Second World War when the 2-day weekend break became standard. There have been subsequent forecasts of a 4-day working week and a 3-day weekend (Poor, 1972). Early finishes on Fridays are now common, but as yet the 5-day working week has remained the most common pattern.

In the late 19th century employers began to recognize annual holidays. At first these were usually holidays without pay for manual workers. Until the First World War paid holidays were non-manual perks in most businesses. Workers in Britain gained a statutory right to at least 2 weeks paid annual holiday only at the outbreak of the Second World War. Public holidays have also increased; Spring, August, May and New Year Bank Holidays have been added to the leisure calendar. Until the 1950s most manual workers had no more than 2 weeks plus public holidays. Subsequently the working year has been the unit of working time that has been trimmed most heavily. In most occupations holiday entitlement has risen to 3, 4 or 5 weeks. The generous holidays that were once considered a privilege of prestigious professions have spread to virtually everyone.

The length of the typical working life has also shrunk. Childhood has been released for schooling. This began when elementary education was made compulsory, up to age 10 initially between 1870 and 1880. Subsequently the statutory school-leaving age has been raised gradually, to 16 on the most recent occasion in 1972. Since then there has been a trend towards staying on for post-compulsory education. And there has been a parallel trend towards earlier retirement. This is despite people remaining healthier and living longer. State retirement pensions first became available in Britain in 1910 and this led to the institutionalization of retirement. Nearly all firms now have retirement ages. It has become accepted that on reaching a given age, usually 65 for men and 60 for women, individuals become entitled to give up work yet retain an income. During their working lives they are deemed to earn the right to pensioned retirement. State pensions alone have never ensured prosperity in retirement, but approximately half of all employees now retire with private, usually occupational pensions in addition to their state incomes. There has been no reduction in the age at which state pensions become available since 1910 but more people have been retiring earlier, many under severance schemes which provide full occupational pensions despite normal service being incomplete.

Since the mid-19th century working time has been assaulted from all sides and the net result is that today people have much more lifetime free from paid employment. The pace of change has not been so rapid that people will have wondered whatever to do with their extra free time. The impact of the long-term historical trend will have been overridden in most people's experience by age effects – moving from being a student to being a worker, and later on into retirement for example – and by the impact of short-term fluctuations which have been liable to create overtime opportunities in one year and short-time working in the next. However, some people have experienced sudden gains in free time at specific points in their lives as standard working hours have taken one of their lurches downwards. This happened when I worked in a bank before entering university in the early 1960s, when Saturday morning opening was abolished (though it has subsequently been re-introduced) and bank employees won a 2-day weekend. Friday nights suddenly became different. So did Saturday afternoons when one felt able to compete at sport on level terms with others who did not have to rush from work. Many employees will have had comparable one-off experiences, perhaps following sudden increases in holiday entitlement, but no one can have experienced the combined impact of all the reductions in normal working time, since they were spread over the last 150 years. However, the fact is that working time accounts for a much smaller proportion of lifetime today than when our great grandparents were in the workforce.

Higher incomes

People are working less than formerly but earning more. This has been possible through improvements in labour productivity sometimes linked to new technology, and otherwise simply to more efficient working practices. Some of the benefits of economic growth have been taken in higher incomes; only a proportion has been taken in the form of increased free time. In fact the growth in personal incomes has been steeper than the growth of free time, from which one might infer that the population has wanted more free time and more money, but rather more of the latter than the former. People today have much more cash available for leisure spending than their parents and grandparents. Since the Second World War standards of living have doubled every 25 years or so. Previously the rate of growth was more modest. Nevertheless, all generations since the 1850s have experienced improvements in their standards of living. Some have experienced temporary cutbacks, during the World Wars for example, but no generation has experienced an overall decline in its prosperity across its entire lifespan. Standards of living have risen so steeply that virtually everyone must have noticed the difference, certainly everyone who has lived this side of the 1930s.

Living standards have risen in all sections of the population. The workforce has benefited from higher wages and salaries. Other groups have benefited from the higher earnings of those on whom they have depended, the state welfare services that have been created during the 20th century, or their own ability to save or borrow in order to spread their personal spending throughout the life course. Many people are still extremely poor relative to average standards of living. This type of poverty, relative poverty, is as widespread and intense in the 1990s as a century ago. And the poorest sections of the population in Britain, the bottom tenth, have experienced no gains in their absolute standards of living during the 1980s and 1990s. But compared with their direct counterparts in earlier times – when workers are compared with workers, and the retired of today with the retired in earlier generations – all groups have become considerably better off.

The feeling of being well off may not have become more widespread. During the 20th century we have learnt how quickly hopes can become expectations, then necessities. Nowadays people feel that they need their motor cars, washing machines, telephones and holidays abroad. Life would become intolerable for many without them. History has confounded some earlier expectations. We now know that once our basic needs have been satisfied we are capable of acquiring an apparently infinite array of wants. It was once believed that once their basic needs were met people would become content, and suffering, striving and conflict would be eradicated. Until the Second World War most occupational groups expressed their aspirations in terms of a 'fair day's pay' and protecting a traditional standard of living. This was in fact people's overriding concern in pre-modern times.

Neither governments nor the people expected living standards to rise progressively from year to year. They hoped to protect their traditional ways of life from the ravages of drought, famine, plague and war. Improvements and declines in a country's fortunes were more likely to be regarded as acts of God than trends over which the people themselves or their rulers had any control. It is different today; governments are judged by their ability to maintain *rising* living standards. Since the Second World War living standards have risen so rapidly that few people will have retained any conception of a normal way of life. The basic standards that earlier generations were anxious to protect are no longer threatened but, rather than everyone feeling comfortable, anxiety over living standards has probably become more widespread and intense. People evaluate their own conditions not against a historical yardstick but by comparing their own circumstances with those of reference groups – people considered comparable to themselves. As everyone becomes better off no one feels more prosperous. There are repeated forecasts that at some point people will see through the illusion of betterment, cease striving for more, and that the engine that has driven industrial societies forward will splutter to a halt, but this does not appear an immediate prospect. Everyone settling for three acres and a cow is not a viable option. A steady state is no longer available. Modern economies either grow or shrink, and likewise people's standards of living. When the communist systems collapsed at the end of the 1980s people who were already poor by Western standards became considerably poorer and one indicator of this was a rise in mortality rates (see Gvozdeva, 1994; Jung, 1990, 1994). Seeking and obtaining improvements in living standards may not be a recipe for contentment but it is probably more satisfying than coping with objective impoverishment.

In Western countries rising standards of living have been reflected in rising levels of leisure activity and spending. Nearly all the participation and ownership graphs slope upwards. The exceptions are invariably cases where older favourites have been replaced by something similar but more attractive such as when radio was superseded by television. We spend more than in the past on what have been regarded as the basics: shelter, food, fuel and clothing. We also spend considerably more on items that were once considered luxuries: overseas holidays, meals out, and domestic entertainment equipment. Leisure has become a major industry, arguably the biggest industry, and it is one of the few economic sectors that is growing virtually everywhere, and growing not only in terms of investment, turnover and profits but employment also.

Home life

The third force, reinforcing the effects of shorter hours of work and higher incomes, fuelling the growth of leisure has been changes in home life. Birth rates and death rates have fallen since the 19th century. The fall in the birth rate had made the small, planned family the norm by the 1930s. By then women were giving birth to far fewer children than their grandparents, only half as many on average, and child rearing had been compressed. This has remained so to the present day. The typical ages of women having their first children fell in the 1950s and 1960s, and have subsequently risen, but the norm of the small, planned family in which all the children are usually born within 10 years, and often much more quickly, has continued. This amounts to a vast change from the time when child-bearing and child rearing were effectively full-time occupations for most adult women. When their own reproductive activities were over the women who survived were once expected to continue using their skills and experience as grandmothers. Child rearing has been lightened, and is now a major occupation for a smaller proportion of most parents' lives than in the past.

The decline in the death rate means that more children survive into adulthood, and more adults into old age. The death of a young child or sibling is no longer a common experience. Similarly, it is now rare for young children to lose their mothers or fathers through death. Divorce and separation are more common than in the 19th century, but in those earlier times families were more likely to be broken by mortality. People who are born fit and healthy can now expect to survive into old age, and older people themselves are living longer. A consequence is that for some people retirement now lasts longer than working life. Most people can now expect a life phase beyond their careers in the labour market during the greater part of which they will be fit, active and mentally alert. The advent of the small, planned family means that many adults' principal responsibilities as parents are now completed well before they retire from employment. Hence the creation of the so-called empty nest life stage. This is a phase in their lives when many workers reach their peak earnings while their domestic responsibilities are becoming less onerous, and a result is that many have time and money for leisure on an unprecedented scale.

A further set of changes in home life has followed the spread of domestic technology. Mains electricity and central heating have made homes more comfortable and have allowed rooms to be put to more uses. Bedrooms today are often used as studies and entertainment centres as well as for sleeping. The numerous gadgets that run on electricity have lightened housework. Deep freezers, refrigerators, cookers, microwave ovens, irons, washing machines, tumble driers, vacuum cleaners and the availability of pre-prepared foods have made it easier to launder clothes, keep dwellings clean, and feed a family. Gadgets that were once science fiction first became luxuries and are now standard domestic equipment.

Simultaneously, homes have become more comfortable with the acquisition of soft furnishings, wall-to-wall carpets, instant heat and illumination, and so on. In addition, dwellings have been equipped with an ever expanding range of leisure equipment; several television sets, radios, video, audio and compact disc players, and home computers have joined or replaced the older card and board games.

All the trends that have generated more leisure are still operating. None has ground to a halt. The containment of working time that began in the 19th century is ongoing. So is economic growth and the rise in standards of living, and likewise the additions to home equipment. The trends are endemic in modern societies. It is difficult to think of reasons why the trends should not continue indefinitely. 'All other relevant things remaining equal' does not appear to be an unwarranted assumption in this case. The growth of leisure is not a recent trend. It is well established, long running and continuing. It is not just a desire to inflate the status of their subject that leads people who work in leisure – providing for other people's, or researching and writing about it – to claim that this sphere of life really is becoming more and more important.

Shaping Modern Leisure

The 19th century

This was the creative period in shaping modern leisure: '. . . there is nothing in the leisure of today that was not visible in 1880' (Cunningham, 1980). All the basic forms were created by then, initially in Britain, the first industrial nation, and these basic forms have remained in place ever since. Work was modernized, life became divided into work and leisure, and the spread of this modern form of time organization was accompanied by the diffusion of clocks and watches. These were not new inventions. The change was that with the modernization of the economy and life in general it became virtually impossible to live without them. During the second half of the 19th century the leisure occasions were also created: at the end of each working day, the weekend and annual holidays. All the subsequent growth has been on these foundations. People now have shorter working days, longer weekends and more holidays, but within recognizable forms. The major leisure provision systems were also created in the 19th century: commercial, voluntary and the public sector. What happened in the 19th century was revolutionary. Everything since, even the developments that some describe as postmodernization, has been evolutionary in comparison.

During the second half of the 19th century the working class became more prosperous. Many families were able to move into the new terraced housing that was built to conform with local authority by-laws. Some were able to buy their homes with the help of the recently created building societies. By the end of the 19th century homes were being connected to mains gas supplies and had fresh water 'on tap'. The availability of fresh drinking water and, after the 1850s, the growing popularity of tea, probably did as much as campaigns for temperance to spread sobriety. The promotion of non-alcoholic beverages, 'soft drinks' including dandelion and burdock, also had some success in replacing beer, especially as thirst quenchers for children and young people. Commerce began to cater for the leisure time and the spending power of the industrial population. During the late 19th century the urban pub was given a style and structure that present-day customers would recognize. The theatre and the music hall developed as alternative places of entertainment. The kinds of entertainment that they offered have subsequently been transferred to radio and screen, but the basic formats were 19th century inventions. The holiday resorts also developed rapidly with their boarding houses, piers and amusement parks. Their expansion was a product of prosperity, railway transport, and employers recognizing holiday weeks (see Walton, 1977; Walvin, 1978). People today take more holidays, and travel to more distant destinations, but the basic ingredients of the annual holiday have remained unchanged.

Modern leisure's voluntary sector also has 19th century origins, and sport was the base for some of the most influential voluntary bodies. Modern sports were first developed in the public schools and universities where they became associated with physical, social and moral well-being. From their very beginning, the development of modern sports – fixing the rules, and organizing competitions – was in the hands of voluntary associations. These bodies were remarkably effective in promoting their sports throughout the country. Association football caught on very quickly and spread as a participant and spectator sport. The sides which competed in the early years of the FA Cup and the Football League were run by committees of amateurs, as indeed were the national bodies. Football clubs that now compete in the Premier League and European competitions were often originally based on schools, colleges, churches and firms. Local teams fostered local pride and gave national reputations to many industrial towns. The 5½-day week, the railways, and the development of a national popular press were all vital to the success of spectator sports, and by the end of the 19th century all the elements of the modern game of football were in place. The leading clubs had professional players, a transfer system was operating, international fixtures were being organized, and crowd misbehaviour had already become an issue (see Dunning *et al.*, 1986; Mason, 1994). Sport was then run mainly on a voluntary basis and so, by the end of the 19th century, were many other leisure activities. Working men's clubs were opening throughout the country. Youth organizations were beginning to develop. And a host of hobbies had their associations of enthusiasts.

The foundations of modern public sector leisure provision were also laid during the second half of the 19th century. A mixture of civic pride and a desire to promote public health and 'worthwhile' uses of free time led towns and cities to build swimming baths, libraries, art galleries, concert halls and museums, and lay out urban parks. Present-day local government departments of leisure and recreation services have evolved directly from these beginnings.

Throughout this period leisure was deeply class divided. The middle and working classes lived in different districts and sent their children to different schools. The classes could usually be recognized instantly by their dress, and they spent their leisure in entirely different places if not in different ways. When Seebohm Rowntree conducted his first study of poverty in York in 1899 he defined the working class (the subjects of his research) as households without servants. Middle class households nearly always had domestic help. This was a common occupation for working class females. Middle class women expected such assistance. Within all social classes there were clear age and sex divisions in ways of life. For example, the new sports were designed by and for men and were played by very few women. The pubs and music halls were adult dominated. Children and young people had to find their own space in the streets.

All the things that people do with their leisure nowadays had close 19th century counterparts. Modern leisure occasions were created in the 19th century and were filled with what are still the modern range of leisure activities. The 19th century was a truly creative period. The subsequent growth of leisure may have been spectacular but it has been less creative. The developments that are currently claimed to be creating a postmodern condition, such as global mass communications, look modest in comparison.

1900–1945

The first half of the 20th century saw two world wars, and the inter-war period is best remembered for its recession and mass unemployment, but overall there was economic growth and a rise in living standards. Even during the recession most people who remained in employment achieved increases in their real incomes. The growth of leisure that began in the 19th century continued during this period.

Local government leisure provision was consolidated. The public baths developed as recreational facilities rather than as places for washing bodies and clothes. Local authorities were made responsible for the development of the Youth Service which addressed the social and recreational needs of 14–21 year olds. The introduction of town and country planning made recreation space an issue within urban areas and on the fringes, and in the countryside.

The voluntary sector developed and was strengthened during the wars which created unmet needs among civilians and during the resettlement of ex-servicemen, while boosting social solidarity. Hobby and interest groups flourished. Working men's clubs became more widespread. Sport continued to develop as a mass participation activity and as mass entertainment. The voluntary youth movements grew to what was to be their peak strength.

Homes became more leisurely especially when dwellings were connected to mains electricity, and a new range of 'white' consumer goods was marketed, most visibly in the department stores that opened in the major cities (see Bowden, 1994). Electric irons were bought by most households between the wars. Refrigerators and washing machines also became available but these were luxury items at the time. The gramophone was another mass marketed commodity, and as ownership spread the popularity of tunes and songs could be measured by sales of records as well as sheet music. However, the domestic gadget which made the greatest difference to home based leisure was undoubtedly the radio. Ownership spread rapidly following the creation of the British Broadcasting Corporation (BBC) in 1926, and the radio became the population's principal source of news and entertainment (see Davies, 1994). Variety acts and drama were transferred from theatres and music halls to the radio which rapidly became a new art form. The radio was a new medium through which performers could become national stars, and musical compositions could become instant 'hits', literally overnight. Television broadcasting began in Britain in 1936 but did not develop a mass audience until after the Second World War.

The other new medium that made stars was the cinema (see Davies, 1994). Between the wars picture palaces were constructed throughout the country. Most adults became cinema-goers, initially to silent then to talking movies, and in the 1930s the typical cinema-goer went more than once a week (see Davies, 1994). The rapid progress of the radio and cinema was linked to a decline in the music hall and live theatre. Alcohol consumption was another casualty. The cinema and radio proved more effective than preaching in reducing time spent in pubs. The decline in alcohol consumption, a long-established pre-modern pastime, was accelerated by the limited licensing hours that were introduced during the First World War, taxation, and restrictions on the number and types of premises able to sell alcohol for in-house or outside consumption.

There were other growth areas in commercial leisure. 'Modern' types of dancing became popular among young people. The dance palais was a place for the young whereas, up to the Second World War, cinemas, pubs and spectator sports attracted mainly adults. Spectator sports widened their appeal and the leading football clubs built new and larger stadiums, many of which remained little changed until the 1990s (see Mason, 1994). Between the wars motoring and flying were pioneered as new leisure activities but these were restricted to the well-to-do (see Howkins and Lowerson, 1979). Cycling was the mass recreation. Rambling was another, made possible by a combination of the spread of prosperity, greater leisure time, rail and bus transport into, and protection of public rights of access in, the countryside. The major holiday resorts expanded as the annual holiday away from home became a normal expectation in working class families.

Throughout this period leisure was becoming more commercial. Much leisure provision was straightforward private enterprise, and some nominally voluntary associations were increasingly operating on more commercial lines. The major spectator sports were leading examples of this. Much the same applied to some public sector provisions. Access to the broadcasts of a public corporation, the BBC, depended on the purchase of radio sets which were produced and marketed by private enterprise, and the 'light' entertainers who gained air time did so on the basis of their mass appeal, and their fees were tailored accordingly.

Post-1945

After the Second World War the growth of leisure resumed. As in earlier decades, economic growth was the principal motor. The three decades following 1945 have been dubbed the '30 glorious years' during which the Western world enjoyed unprecedented sustained economic growth and full employment. Since then economic expansion has continued, but at a slower pace, and some sections of the population, mainly outside the paid workforce, have not shared in the growth in incomes. Even so, the overall growth of leisure has continued.

In the immediate post-war years the most visible beneficiaries of the new economic conditions were young people. The strong demand for labour eliminated youth unemployment in most regions, and drew teenagers' earnings towards adult levels. Hence the creation of affluent young workers who became prime targets for a number of consumer industries, and a series of flamboyant youth styles and fashions was a high profile outcome (Abrams, 1961). Subsequently other sections of the population have joined the consumer society.

The commercialization of leisure, involving the shaping of more leisure goods and services into commodities which can be bought and sold, has continued, accompanied by intellectuals' critiques of consumer culture and 'commodity fetishism' which, some argue, create wants which differ from people's real needs. These arguments are considered fully in Chapter 7. It is certainly true that leisure has become big business, and the major voluntary organizations increasingly operate on commercial lines, in running participant as well as spectator sport facilities for example. It is also true that some older voluntary organizations have seen their roles diminish as their former members have turned to commercial alternatives. The traditional youth organizations have been prime examples of this. Some nominally voluntary bodies now depend heavily on state support; youth organizations and amateur sports associations for instance. However, it is still the case that most participant sport is run on a genuinely voluntary basis, and virtually every leisure interest has enthusiasts who organize themselves in clubs and societies (see Bishop and Hoggett, 1986).

There is also a substantial public sector which has grown rather than contracted alongside the development of commercial leisure since 1945. Since the 1970s most local authorities have rationalized the management of their leisure services into single departments and have sought overarching leisure policies. Central government has become more prominent in leisure provision. National councils and commissions with responsibilities for sport, the arts, the countryside and tourism have been created. Since 1992 central government's leisure responsibilities have been drawn together in a Department of Culture, Media and Sport, and since 1994 the National Lottery has become a major source of public funding. The 'good causes' that benefit from the National Lottery are sport, the arts, heritage, charities, and the Millennium Fund which has a leisure dimension.

In some respects public provisions have become more commercial. Public providers have been encouraged to operate on business lines. Since 1992 local authorities have been obliged to submit the management of most of their leisure services to competitive tendering. But it remains the case that many public leisure provisions are subsidized from the National Lottery or from general central or local government budgets, thereby sustaining facilities that could not survive on a purely commercial basis or on voluntary effort. Participant sport and high culture would both be substantially reduced if all public subsidies were withdrawn.

Rising standards of living have enabled households to acquire more leisure equipment, and since 1945 one of the most significant of these acquisitions has been the television. The television set became a standard lounge fixture in the 1950s and ever since it has accounted for a substantial slice of leisure time. Radio and cinema audiences were decimated as television soared in popularity. These former leading media have subsequently adjusted by moving into market niches that television is less able to fill. Radio has targeted the audiences for various types of music, and car drivers more generally. The cinema has adjusted by producing specialist products to be watched at multi-screen complexes by specific taste publics, mostly composed of young people rather than adults who are more difficult to tempt from their homes and televisions. The press has also been affected by the rise of television. Sales have been maintained by newspapers either offering more detailed news coverage and analysis than other media (the quality broadsheets) or entertainment and, in particular, guides to other media, especially television programmes, and stories about media personalities (the tabloids). Most households now contain more than one television receiver and additional items which widen the choice of screen entertainment – videos, camcorders, cable and satellite links. Most homes also have a range of audio equipment to play records, tapes and compact discs. The improved quality of the furnishings has further enhanced the home as a leisure environment, as have central heating and telephones. The manner in which the telephone enables conversation to be purchased, and the cost compared with a video rental or a glass of beer, is an excellent example of commodification.

Despite the appeal of television, since 1945 leisure has not become more home based. During the 1960s car ownership spread to the majority of households and made it easier for people (except those without cars) to go out. Cinema audiences have shrunk to a small fraction of their sizes in the 1930s and 1940s, and the film making industry has adjusted by designing more of its products for video and television distribution. The crowds at sports events have also shrunk and top sport is increasingly marketed to the television viewing public (see Whannel, 1986). But there has been no decline since the 1950s in the audiences for live theatre and music, or visits to art galleries, stately homes, coastal resorts, countryside honeyspots and other venues for trips out. The frequency with which people go out has made catering into a leading leisure market. People have been going out more frequently for meals, sometimes as the main activity and sometimes while travelling to or from other leisure destinations. Alcohol consumption has also reversed its former long-term decline. Between the 1950s and 1980s per capita consumption doubled (see Goddard, 1991). Like food, drink can either be the focus of a leisure occasion or an incidental alongside other activities. The places where alcohol is available for consumption on the premises, and for sale for consumption outside, have multiplied. Alongside these trends drinking has become socially acceptable, and often expected, within sections of the population (women and young people for example) where it was once uncommon. By age 16 most young people are now drinking regularly, and the vast majority of their parents know about and condone their children's conduct (Sharp *et al.*, 1988).

The holiday industry has been another beneficiary of mass prosperity. Since the Second World War the holiday away from home has been an event in most people's annual leisure calendars, and since the 1950s the number and proportion of holidays taken abroad have risen. Meanwhile the UK has been attracting more tourists from overseas. The domestic population has been travelling further, to more exotic destinations, on main holidays, but it has also been taking more second holidays and weekend breaks most of which do not involve overseas travel. Camping holidays with tents and caravans have risen in popularity. It is this all round growth that has made tourism into a major, arguably the biggest single, industry in the world.

The growth in holidays away and going out more generally have expanded the markets for leisure goods and clothing. Sports goods have become big business and sportswear even more so as the boundary between sports clothing and general leisure attire has dissolved. The tendency nowadays is not for people to 'dress up' but to dress casually when they go out which is just one of many indications of how life in general has become more leisurely.

Long-term Trends

Commercialization

Within the overall growth, and within the overall shape created in the 19th century, it is possible to identify several long-term trends in how people spend their leisure. One of these, commercialization, has already been highlighted. It is important to remember that this has been a trend, but remains a far from absolute condition. Leisure has not been totally commercialized. It is simply more commercial than at the turn of the century and the basic reason for this is that people have more money.

Blurred divisions

A second long-term trend has been a blurring of formerly clearer divisions in leisure tastes and activities by age, sex, social class and region (Howkins and Lowerson, 1979; see Davies, 1992; Mason, 1994). There are numerous reasons for this blurring, many rooted outside leisure itself, but the development of national, often international, leisure markets has played a part, as has commodification. Private enterprise in general, though not every single merchandiser, has a vested interest in gaining access to the widest possible markets. Particular products, musical genres for instance, may be targeted at specific groups. Some products, holiday resorts for example, may cultivate exclusive reputations. But the tourist industry in general, and many resorts, can be relied on to seek customers from all age groups and social classes. As the population has grown more prosperous, more and more activities that were once confined to the well-to-do have become mass pastimes. Motoring and eating out are examples. Simultaneously, goods and services that were once considered coarse by groups who prided themselves on their more refined tastes have spread upmarket. People from all social classes now listen to popular music (see Longhurst, 1996). Classical fare is enjoyed as well as, not instead of, pop.

The blurring of divisions will have diffused leisure's bonding function. The blurring does not mean that everyone now does everything. Leisure has not been standardized. Most leisure activities are minority interests which draw together their particular followers and separate them from other members of the public. Social divisions become blurred when interests attract participants from many age groups, regions and social classes, and from both sexes. So the various taste publics (people who share particular interests) in which individuals are involved all tend to have socially diverse memberships, and each of the many activities in which any person is involved creates a different set of peers. When this happens the leisure networks in which people operate cease to be socially superimposed; working class people going to particular holiday resorts, wearing distinctive clothing, listening to 'their' types of music and so on. It thereby becomes more difficult to judge any individual's occupational class, age, sex or home region from a knowledge of what he or she does when at leisure.

Individualization

This trend is related to the blurring described above. It occurs as leisure ceases to separate the population clearly into distinct groups. Again, it is important to stress that individualization is a trend rather than an absolute state. In this case the trend is towards every individual having a particular combination of leisure interests and activities, and a unique leisure career. Individuals develop personal stocks of leisure skills and interests, some of which may be drawn from, and others may subsequently be used in, other spheres of life. No skill or interest will be unique to any person, just the total package. Exactly how far these trends have gone, and their sources, are explored in detail in Chapters 3, 4 and 5 which deal with the relationships between leisure activities and individuals' occupations, gender and age groups.

Pacification

A fourth long-term trend has been pacification. Eric Dunning has argued that many developments in leisure are best understood as part of a broader civilizing process (see Dunning *et al.*, 1992; Dunning and Rojek, 1992). It is said that the development of modern societies has seen life become increasingly civil, peaceful and orderly. Sport is said to illustrate the civilizing process in two ways. First, much aggression has been taken out of society and contained within sports. Second, sports themselves have become less violent. Hacking is no longer an acceptable tactic in football, and nowadays amateur boxers are fitted with headguards. It can be argued that the concern about soccer hooliganism that mounted from the 1960s to the 1980s was not so much a response to the behaviour of spectators deteriorating, but the wider society, and the majority of those who wanted to watch sport, becoming less tolerant of such conduct. Pacification is evident in many areas of leisure. Playground fighting is no longer an acceptable schoolboy recreation. Street fights after closing time are no longer simply accepted as part of weekend life. And it is not only leisure that has become more civilized. Schoolteachers are no longer permitted to administer corporal punishment and its use within families has become controversial.

Reservations and Qualifications

All the above trends, like the growth of leisure, have been very long term. No one seriously questions that the trends have occurred, but in recent years a number of reservations have been expressed. These concern the extent to which leisure has grown, whether all sections of the population have been affected in the same way, whether people today really have historically unprecedented amounts of leisure, the type of leisure that is now playing a larger part in people's lives, and whether the upward trend is still in process and can be expected to continue indefinitely.

More leisure for everyone or just for some?

One of the underlying forces behind the growth of leisure has been the containment of work, and this historical story is often told as if it applied to all sections of the population. In fact it is male history. The history of women's lives during the last 150 years reads very differently. Presenting a male history of work and leisure time as if it was gender neutral is one example of a gender blindness that has often left women hidden from view (see Chapter 4). During the latter part of the 19th century the main trend among working class married women in Britain was not a steady reduction in their hours of paid employment but their wholesale removal from the workforce (see Hakim, 1993). At that time upper and middle class wives did not work outside their homes (or in their homes in some cases) and the bourgeois ideal was spread downwards. The woman as homemaker was promoted as an ideal. She was to provide a comfortable and nurturant base for children and male employees to return to at the end of each working day. Even with the benefit of hindsight there are very different interpretations of this trend. Some see it as a re-assertion of patriarchy, disempowering women and forcing them into economic servitude and therefore social and emotional dependence on men. Others argue that women benefited when employers accepted that men should be paid a 'family wage' thereby making it unnecessary for wives and mothers to seek employment in the mines and factories.

By the beginning of the 20th century unmarried females, mostly young women, were the only female group most of whom were in employment. During both World Wars single and married women were drawn into the workforce to do jobs vacated by male servicemen, and to staff the munitions factories and other war industries. On both occasions, following the cessation of hostilities, most of these women returned to domesticity and their jobs were re-occupied by men. However, since the 1940s there has been a steady rise in the proportion of married women in employment. Up to now most of these women have taken part-time jobs, and have returned to the labour market only when their youngest children have become school aged, but since the 1970s women have been shortening their career breaks. More have been taking maternity leave instead of terminating their employment, and by the end of the 1980s the number of women in full-time employment was rising steadily (see Hakim, 1993). Rather than women's hours of paid work steadily declining, there has been an overall increase during the last 50 years. There are many women today whose total work weeks, when their paid employment and unpaid domestic work are totalled, exceed 70 hours. Such long work schedules have not been buried in 19th century history. Of course, women have been affected, more than men, by the decline in fertility and the spread of domestic labour saving technology. But the fact remains that one of the major strands in the story about the historical growth of leisure bears no resemblance to the experiences of women.

In general men's hours of work have declined, but this has not applied among males in all occupations. In the professions and among the self-employed there is no evidence of any overall decline in hours of work since the 19th century. Indeed, it is only during the 20th century that some professions have been transformed from bases for gentlemanly ways of life into modern occupations. Solicitors, barristers, accountants, bank managers, teachers, civil servants and local government officers have not experienced any reductions in their daily or weekly hours of work, or increases in their holiday entitlement, or reductions in the ages at which they have guaranteed retirement rights on full pensions. As the next chapter explains, in recent years working time in many jobs such as these has actually risen.

The story about the working day, week, year and life being gradually scaled down is in fact the history of male industrial manual workers. At the beginning of the 20th century the manual grades accounted for around fourth-fifths of the entire workforce. At that time the occupational class structure was bottom heavy and it may have been reasonable to present the stories of manual workers as the history of 'the people', but today less than half of male and female employees are in manual jobs. Employment has shifted towards occupations in which working time has not been in long-term decline and where it is currently tending to increase rather than decrease.

Since it bears little resemblance to women's history, and does not apply in a substantial number of male occupations, is it legitimate to refer to the containment of work as a *general* historical trend? This question cannot be sidelined by re-asserting that the typical worker today does in fact have a shorter working day, week, year and life than his (usually) 19th century counterpart.

Does less work mean more leisure?

During the 100 years following 1870 some of the time released from paid work was used to give employees shorter working days and weeks, and longer holidays, but a greater quantity of the 'new' work-free lifetime increased the size of groups outside the paid workforce, mainly school-children and the retired (see Wilenski, 1963). Since the children who were released from employment in the mines, factories and shops were expected to attend school, this product of the containment of working time can hardly be construed as a growth of leisure. Although the case is less clear cut, question marks can be placed against the leisure implications of the expansion of the time spent in retirement (see Parker, 1979). Many of the retired have lived, and continue to subsist, on very low incomes, much lower than when they were in employment. Their reduced incomes inevitably reduce their leisure opportunities and will have done so more sharply as leisure has become more commercial. Is retirement really leisure time or is much of it literally spare time? This issue is taken up in Chapter 5. Such questions have become increasingly pertinent since the 1960s because virtually all the recent additions to work-free time have involved young people delaying their entry into employment, people retiring younger and living longer, plus the increase in the numbers unemployed. Whether there are any senses in which people who move from jobs into unemployment can be said to experience an increase in their leisure is discussed in Chapter 3. For present purposes it is sufficient to introduce serious reservations against any presumption that lifetime released from paid work must have led to an equivalent growth in leisure.

The time that people spend travelling to work is relevant here. Employees who have experienced reductions in the hours when they are at work may spend many hours per week travelling to and from their jobs. Long distance commuting by road and rail has become common in south-east England. Working days of over 12 hours have remained common when travelling time is included.

There have been comparable trends limiting the growth of leisure that would have followed the compression of child rearing and the acquisition of domestic technology if all other things had remained equal. In the event standards of home and child care have risen. Many parents no longer consider it safe to allow their children to play out without supervision or to travel to school or leisure unaccompanied. Between 1971 and 1991 the proportion of Britain's 7 year olds who normally went to school on their own declined from 70% to just 7% (Hillman, 1991). Washing machines have taken much of the labour out of laundering but have also led to households doing more washing. It is now common for garments to be washed after being worn just once. Electric cleaners have made it easier to remove dust from carpets but nowadays this chore is likely to be performed several times a week. Meanwhile, as home and car ownership have become more widespread, more time has been devoted to odd jobs on the owners' dwellings, cars and gardens.

The historical benchmark

Leisure is most likely to appear to be on a long-term upward course when the early industrial period is adopted as the starting point for history. There were no time budget studies in pre-modern times but we know that in the countryside the burden of work was not consistently oppressive. People worked by 'natural rhythms'. At certain times such as harvests there was much that had to be done quickly but this did not apply week in and week out throughout the year, and the calendar was sprinkled with saints' and feast days. During the take-off into industrialism working time was increased in the face of considerable workforce resistance, and it has been argued that since the 19th century the containment of working time has achieved no more than a recovery of the free time that was lost during the previous century (Wilenski, 1963). With a longer historical perspective, the early stages of industrialism stand out as the exceptional period when the demands of work were raised to a peak. Since then, it can be argued, the free time that was lost has been reclaimed and working time has returned to its historically normal level, but, even in the most economically advanced countries, people today are not enjoying more free time than in any earlier era, nor compared with many present-day peoples whose ways of life remain pre-modern.

The declining pace of the decline in working time

In Britain the steepest decline in working time occurred between 1850 and the Second World War. Since then there have been reductions in the number of weeks worked per year, and in the number of years in the typical (male) working life, but little change in daily or weekly working hours. Benjamin Hunnicutt (1988) has argued that around the time of the Second World War the labour movements in Western countries changed their objectives. Formerly, he argues, they were inspired by a vision of a society in which people would work less and more time would be their own. Subsequently organized labour has demanded full employment, and the right to earn and spend.

Since the 1970s some countries, including Britain, have recorded net increases in the weekly hours worked by full-time employees (see Fajertag, 1996; Holliday, 1996). Average hours of work have continued to decline as the number of part-time jobs, filled mainly by women, has risen, but if these occupations are excluded from, or properly taken into account in the calculations, the hours worked by both men and women in most occupations are seen to have either increased or to have settled on a plateau (see Zuzanek *et al.*, 1998). Since 1981 there has been no decline in the total number of hours spent in paid employment in Britain, and only a very slight decrease in the annual working time of the typical full-time (FT) worker (see Table 2.1).

Part of the explanation lies in the shift of employment towards management and professional jobs, and self-employment, in which there has been no long-term decline in working time. Rather, within these and some manual occupations people in Britain have actually begun working longer. This trend is discussed fully in Chapter 3. It may prove temporary. A long-term increase in working time seems improbable. It seems more likely that having recovered the time that was lost during the take-off into industrialism, working time has returned to the historical plateau that human beings, maybe as a result of their earlier biological evolution, experience as normal and no longer feel a need for further reductions. It is also likely that, with the progressive commercialization of leisure, workers' desire for higher earnings will have taken precedence over their interest in still more free time. These possibilities are considered fully in the following chapter. At present it is sufficient to reinforce the grounds for doubting whether we really live in an era in which leisure has been and is still growing, and will continue to do so remorselessly in the absence of unforeseen developments.

However, it could be equally unwise to assume that the former decline in working time has ended permanently. Historians have shown that reductions in hours of work, when they occurred, were not simply delivered by economic progress but had to be won through collective bargaining (Bienefeld, 1972; Rosenweig, 1983; Jones, 1986). If organized labour's will to achieve still shorter hours has weakened, or if broader trends have undermined the power of organized labour, there seems little reason to predict further reductions in working time, but this is a topic on which prediction is hazardous. It is possible that, before long, organized labour may adopt the view that the only way to prevent an otherwise inexorable spread of mass unemployment is to reduce working time. In some European countries the trade unions have not only adopted this view but have persuaded their

Table 2.1. Hours worked in Britain.

	Total hours of work (billions)	Annual hours worked per FT worker
1971	45.5	1999
1981	40.5	1844
1991	42.1	1829
1996	40.9	1832

Source: Martin and Mason (1998).

'social partners', governments and employers, that normal working hours must be trimmed in the interests of social cohesion (see Blyton and Trinczek, 1996). Up to now this has not led to a substantial all-round drop in working time in any Western country, but it is impossible to rule out a future downward lurch.

The quality of the new leisure

A final set of reservations concerns the quality of any new leisure that has been created. First, people may be spending more on their holidays, by going abroad instead of to domestic resorts for example, but does this make their holidays better? Is it misleading to assume that leisure can be weighed by the amounts of money that people spend?

Second, some of the apparent growth arises from the commodification of activities that were formerly self-organized. Music in the home today is most likely to be from tapes, the radio or compact discs. People undoubtedly spend more on music, and possess more leisure equipment which allows the music to be heard, than in the days when families made their own music and sung along, but can this be equated with growth?

Third, Fred Hirsch (1977) has drawn attention to the fact that there is a 'positional element' in much personal consumption, meaning that the satisfactions that people derive do not depend wholly on their being able to have or do something, but also on other people being excluded. The satisfactions that arise from driving the latest model motor car, or being able to wear the most fashionable clothes, or to go on exclusive holidays are examples. Hirsch argues that as standards of living rise the positional element in consumption grows in importance, and the end result can be likened to a crowd where everyone stands on tiptoe but no one sees further. Everyone may be doing and spending more yet no one may be more satisfied than formerly.

A fourth argument draws attention to the spread of time pressure. Staffen Linder noted in 1970 that leisure time was growing more slowly than personal incomes and spending, and the end result, he argued, would be a 'harried leisure class'. Rather than the growth of leisure leading to more relaxed ways of life he forecast that people would encounter increasing difficulties in 'finding the time'. There certainly seems to have been a trend in this direction judging by the popularity of fast foods and fast sports such as squash rather than cricket matches which can last all day. Geof Godbey (1975) has argued that rather than a long-term growth of leisure, the condition which has recently spread is better described as 'anti-leisure'. This set of arguments does not dispute that overall people today have more leisure time, spend more and engage in more leisure activities than formerly, but simply points out that the end result may not be exactly what would have been expected given the ordinary meanings of leisure. Godbey claims that in fact life has become less leisurely.

Current Trends and Future Prospects

The qualifications and reservations do not completely demolish the historical tale of leisure's growth. The fact remains that work-free time, leisure spending and leisure activity have grown and are continuing to grow, and the following chapters confirm that social divisions have been and are still being blurred by these trends, that individuals are developing and expressing more personal interests through leisure, and doing so in more civilized ways than a century ago. Nevertheless, in recent years a new set of additional trends has operated.

The distribution of leisure

Time, spending power and levels of activity have continued to grow, as rapidly as ever before, but these different components of leisure have been distributed to entirely different sections of the population. Instead of most people experiencing marginal gains in leisure time, spending power and levels of activity, the population appears to be dividing into separate leisure classes. One class is working longer, earning more and can afford to participate in more activities but is experiencing mounting time pressure. A second class is time rich but money poor and has spare time rather than leisure as commonly understood. It is also possible to distinguish a third leisure class composed of people who work in the leisure and other consumer industries, generally in low status, low paid, seasonal or otherwise temporary, often part-time jobs which require unsocial hours of work. Many of these jobs require 'aesthetic labour' in which the employees' appearance and demeanour must be to the customers' satisfaction (see Nickson *et al.*, 1998; Tyler and Abbott, 1998). There is more work to be found in leisure nowadays but the downside of this picture is that most of the jobs are poor quality. There may be more jobs than formerly in the management and professional grades. In this sense employment is being upgraded, but this process is not affecting all occupations. There is still plenty of menial work in the world's most prosperous countries, and much of this work is in leisure (see Seabrook, 1988). The Victorian class of domestic servants has not been socially promoted so much as transferred into the new leisure industries.

Optimists and pessimists

It is possible to construct a highly pessimistic picture of the direction in which the so-called leisure society is heading. Forecasts of a leisure society used to be optimistic. They anticipated that life would become less pressured, and that people would be able to do more things for the sheer enjoyment (see Box 2.1).

Box 2.1. The Inglehart thesis (1997).

Richard Inglehart claims that a silent revolution is in process throughout the Western world: a decline in materialist, and the spread of post-materialist values.

People today are said to be less interested than in the past in gaining higher and higher incomes and more and more goods. We are said to be becoming more concerned with 'quality' – in our experiences, relationships and time. Politics that appeal to people's material self-interests are said to be in decline. New social movements concerned with the environment and civil rights are said to be examples of the influence of post-materialist values.

Inglehart's own research, in several different countries, suggests that successive cohorts of young people have been less materialist than their predecessors. According to Inglehart, economic security in childhood and youth is the crucial condition leading to a decline of materialism.

However, other researchers' evidence indicates that the trends may be more complicated. For example, Helena Helve's (1998) studies of young people in Finland suggests that the same individuals are quite capable of subscribing to both materialist and so-called post-materialist values. For example, it is possible to support equality between the sexes and, simultaneously, to be interested in making plenty of money.

Much of the recent writing on trends in leisure expresses disillusionment and cynicism rather than optimism (see in particular Seabrook, 1988). One point of agreement between the older story of leisure's remorseless growth and the 'revised version' is their regard for the values fostered by leisure as an important ingredient in the 'cement' that locks the population into the present-day socio-economic system. However, the revised version warns that the cement may eventually crumble. Those who are able to do so are tending to work as long and as hard as possible so as to earn and spend as much as they can, but at some point their motivation could slacken. Fred Hirsch (1977), who drew attention to the positional element in personal consumption has insisted that there are social as well as physical limits to economic growth. At some point he believes that people will tire of standing on tiptoe and seeing no further. Frustration is at least equally likely among those who are time rich but whose consumer aspirations are thwarted. Some argue that recent trends simply cannot continue indefinitely and that leisure research and theory should be seeking routes out of the impending impasse.

Time pioneers?

In Germany, Karl Horning and his colleagues (1995) studied a small group of employees who they describe as 'time pioneers' who had voluntarily reduced their weekly hours of work to 20–32. No claim is made that these time pioneers represent a trend. They were contacted through publicity in national and regional newspapers and radio programmes. Among the 36 cases studied, half were considered 'time conventionalists'. They had reduced their hours of work for a specific reason such as becoming a single parent, and had not challenged normal time structures but had filled their additional non-working time with housework, child care or leisure activities.

The true time pioneers, in contrast, had changed their ways of life because they wanted more time for themselves rather than to do anything in particular. Their earnings from the shortened work-weeks were between 1000 and 3000 Deutschmarks (roughly £450–1350) per month. They had calculated how much money they needed to support a basic lifestyle and had learnt to live within it. For some this had not been difficult; their lifestyles had always been frugal and their aspirations non-materialistic. The others had revised their consumer habits and had consciously rejected goods-intensive lifestyles. All the time pioneers valued their temporal affluence and treated this as a key criterion of well-being. They appreciated life's slower pace and leisureliness, being able to pace and plan their own time, and being able to create time buffers so as to prevent pressure building up.

According to their own reports, the time pioneers were regarded as odd by their work colleagues. They were in fact highly committed to their work but to the tasks rather than the status, career prospects, income maximization and other aspects of the general employment culture. This was probably why they were experiencing mixtures of resentment, envy, confrontation and aggression from colleagues and supervisors. Nevertheless, all the time pioneers stressed their good fortune. They regarded themselves as privileged. They had experienced full-time work as too stressful, exacting and costly, and spoke of how their changed schedules had transformed their lives; how they had started to live and taken a qualitative leap forward. Having taken this leap none wanted to turn back.

Horning and his colleagues realized that their time pioneers were exceptional but demanded that they be recognized as people who were pioneering a radically different, and viable way of living. There are no signs as yet of a widespread demand for greatly reduced hours of work and incomes, but there are signs of people who have 'made it' in terms of earnings and status querying the meaning of success. Ray Pahl (1995) studied ten such people in Britain, eight men and two women. One person had made his money and cut-back to an extremely short working week. Others were continuing on arduous work schedules. But in some way or another they were all confronting problems in managing their commitments and investments in economic and personal life, and in deciding exactly what success was. These issues reappear in the following chapter on work and leisure, and in Chapter 7 on consumerism and consumer culture.

Chapter 3

Work and Leisure

This couplet has always been prominent in the sociology of leisure. The work–leisure relationship was in fact among the subject's original issues. In Britain the sociology of leisure emerged in the 1960s largely as an offshoot from the sociology of work. When sociologists then studied leisure the chances were that their core interests really lay in the long arm of the job.

Researchers have emphasized that many features of individuals' jobs can influence their leisure. Hours of work and pay are obvious examples. The containment of working time has always been recognized as one of the long-term trends responsible for the growth of leisure. Rising incomes have had similar recognition. Indeed, pay and hours of work have been responsible for some of the clearest and most consistent findings in work–leisure research. From the beginnings of such enquiries it has been found repeatedly that the higher strata do more of most things, the main exception being television viewing (see for example Havighurst and Feigenbaum, 1959; Dawson, 1988a, 1988b). Perhaps this particular work–leisure relationship has never received the attention, measured by the number of dedicated enquiries or pages of reporting, that its importance merits. Part of the explanation is that researchers cannot repeatedly earn acclaim for the rediscovery, and there are limited ways of saying, that the higher socio-economic strata do more. One reason for the relationship is quite obvious – money – and there are no prizes for repeatedly confirming the obvious.

An upshot is that researchers have always stressed that the implications of the pay people receive and the number of hours they work do not exhaust work's leisure effects. In the 1950s and 1960s they were equally if not more concerned to investigate how work-based social relationships, interests, and states of mind could pervade people's leisure. There was also sustained interest in the leisure implications of not just how many hours people worked but also how these hours were scheduled, especially whether people worked standard hours or on a shift system.

Subsequently, in the late 1970s and 1980s, these entire streams of research encountered criticism. In focusing upon paid work the research was accused of concentrating on men's work and leisure, and neglecting the leisure implications of other types of work such as housework and caring, typically unpaid, and performed mainly by women. During the 1980s gender challenged work as the leading issue in the study of leisure. A consequence is that there is little danger today of anyone neglecting the implications of the unpaid work that is done mainly by women, and to a lesser extent by men. The implications of unpaid work are considered in detail in Chapter 4 which deals with leisure and gender while the present chapter concentrates on the implications of paid employment. This continues to merit detailed attention if only because it accounts for a substantial portion of most people's time – women's as well as men's nowadays because by the mid-1990s women made up just over 49% of all paid workers in Britain, and in many regions the majority of the workforce was female (Noon and Blyton, 1997). Today most men and women are in paid jobs for the greater part of their adult lives, and work supplies most of the income that people spend on leisure activities.

Work–leisure research is currently emerging from a hiatus. The slackening of research in the 1980s was not because the relationship was no longer considered important. Nor had its investigation fallen out of fashion. Rather, the hiatus was due to the work–leisure issues of the 1960s and 1970s receding in importance, partly because the questions of that era were answered satisfactorily, but also because work was changing and the research issues needed to be redefined. Former social and psychological work–leisure links were being undermined by broader social and economic trends. The problems of working non-standard hours were being transformed by the spread of flexibility throughout the economy and labour market. Another change is that research into work and leisure until the 1970s was conducted in a context where hours of work were in long-term decline and there was a corresponding growth in leisure time. Subsequently, as the previous chapter explained, many of Britain's full-time employees (and those in some other relatively prosperous countries) have begun working longer on average and their leisure time has contracted. A further change is that the significance of being employed in different kinds of occupations has been partly eclipsed by the spread of unemployment and the division between people in jobs of all types and the unemployed.

Spillover and Compensation

The long arm of the job

In the 1960s and 1970s there was much research, and even more debate, about the socio-psychological processes whereby the long arm of the job governed people's leisure. Investigators paid attention to how particular occupations induced characteristic states of mind, surrounded workers with work-based social relationships, and gave them interests and social identities, and a social standing, which they carried into the rest of their lives. Just as at a macro-level leisure time had been created by the modern organization of paid work, so, it was argued, at a micro-level the particular kinds of employment that individuals practised shaped their particular orientations towards, and uses of, leisure.

Two main kinds of work–leisure relationship were identified, usually from a mixture of anecdotal and case study evidence. The first was *spillover* where work-based relationships and interests, and social and technical skills, spread into leisure. A common example given was white-collar workers who were able to use occupational skills in running voluntary associations. A less attractive example of spillover was said to be the workers with routine, mundane, mind-numbing jobs whose mentalities were so stunted that they were content to spend most of their leisure being passively entertained. The second type of work–leisure relationship was *compensation* where individuals used their free time to seek experiences which they could not obtain at work. An example frequently offered was the desk bound executive who played sport in the evenings and at weekends. Another was individuals who were denied opportunities to display initiative at work who used their leisure to demonstrate their autonomy ostentatiously.

Opposition, extension and neutrality

It proved embarrassingly easy to find a theory – spillover or compensation – to fit virtually every case, and an obvious problem was to explain what triggered the kind of influence that would operate. A major intervention in this debate was by Stanley Parker (1971, 1983) who affirmed the need for a 'holistic' approach in which work and leisure were viewed as a totality, and proposed that whether spillover or compensation operated depended mainly on the nature of the job. Parker argued that compensation, which he retitled the *opposition pattern*, was most likely when people's jobs were experienced as hard and painful and aroused negative feelings. Under these circumstances Parker suggested that individuals would tend to draw a clear boundary between their work and their leisure. They would look forward to the end of each working day and week. On leaving the workplace they would experience feelings of release and freedom, and would celebrate the opportunity to express their true selves. Parker did not actually study a group of workers in which this opposition pattern prevailed, but he was able to refer to evidence from other studies of occupational groups such as deep sea trawlermen who came onshore and indulged in binges during which they enjoyed drink and female company, and sought hedonistic pleasures, which were unavailable at sea.

Parker argued that spillover, which he retitled the *extension pattern*, tended to occur when people liked their jobs and became absorbed in their work, when occupations fostered amicable and cooperative relationships between colleagues, and when individuals were required to represent their professions and organizations to the wider public. Under these circumstances Parker argued that employees would sometimes find it difficult to decide where their work ended and their leisure commenced. They would spend some of their free time reading professional literature and thinking about work tasks, often in the company of workplace colleagues. Parker found that this type of leisure was more common among the youth employment officers and child care officers that he studied (the predecessors of present-day careers guidance officers and social workers) than among a sample of bank clerks.

Between extension and opposition Parker identified a midway *neutrality pattern* which seemed to apply to many of the bank clerks. He argued that when people were neither enthusiastic about, nor hated their jobs, they were likely to treat their work and leisure as just different, complementary parts of their lives. Work would not extend into leisure, but neither would individuals feel a need to compensate for the deprivations endured each working day.

Criticisms

A problem that soon became apparent with Parker's typology and explanations was that there were too many exceptions. Work might extend into leisure because individuals were highly involved in their jobs and identified with their occupations. But it was noted that occupational communities, which carried workplace social relationships and interests into leisure time, could also arise because the nature or locations of their jobs separated employees from the wider community, as has been observed among coal miners, the police, prison officers and railwaymen (see for example Salaman, 1974). It was also apparent that people who liked their jobs could still use some of their leisure to gain complementary experiences, like exercise-seeking executives.

It was also noted that Parker's theory was not explaining what people did with their leisure so much as who they spent their time with, and how they felt about their leisure activities and relationships. It was observed that some leisure interests were popular among groups with very different occupations. For example, Mott (1973) noted that pigeon fancying was popular among some groups of skilled weavers and in some coal-mining communities. The explanation could have been that one group (the miners) was compensating and that spillover was operating among the weavers. Alternatively, both groups could have been enjoying the activity for much the same reasons and deriving similar satisfactions.

A third problem was that work–leisure relationships which could be illustrated persuasively in case studies proved difficult to identify in sample surveys which covered people in a variety of occupations. In the data sets which such studies produced it was sometimes difficult to discern any powerful spillover or compensatory relationships. Such studies have found rather more evidence of spillover than compensation but all the relationships discovered, between interest in work and uses of leisure for example, have been rather weak (Champoux, 1978; Miller and Kohn, 1983; Zuzanek and Mannell, 1983). Some uses of leisure are found to be popular in virtually all occupational groups: watching television, taking holidays and visiting family members for example. A study of coal-miners, steelworkers, factory employees and white-collar staff in South Wales in the early 1980s found common features such as these alongside clear age and sex differences in uses of leisure within all the occupational groups, but found no distinctive occupational patterns except among the coal-miners (Lloyd, 1986).

Occupational change

This entire genre of research has subsequently become almost extinct. Perhaps the theorists were always mistaken or overstated the strength of spillover and compensatory relationships. However, it is likely that such relationships as may have existed in the past will have weakened. The distinctive work–leisure relationships that were noted in case studies tended to be in occupations practised mainly by one sex (invariably men), usually for their entire working lives, and were located in towns and villages dominated by major industries where workmates tended to be neighbours, and work and community influences on leisure were reinforcing. Recent economic and social trends have broken up such social formations. There are now fewer towns and cities dominated by single firms and industries. The pace of economic and technological change means that occupations can no longer be relied on to last for an entire working life. The private motor car enables firms' workforces to be drawn from wider catchment areas than in the past, and the residents in most districts are now employed in many different occupations. Trends towards home-centred lifestyles will also have weakened former work effects; so will affluence which gives some individuals and households wider choices over what to buy and what to do whether in their homes or outside.

This does not mean that work will no longer have important leisure effects; rather that the types of effects will have changed due to changes in the nature of both work and leisure. Spillover and compensation have probably become less powerful. As we shall see, much the same applies (being overtaken by changes in the nature of work) to another, once substantial, stream of research into the implications of different work schedules.

Shift Work

The development of shift systems

Shift systems are virtually as old as modern industry. 19th century manufacturers experimented with shifts as a way of maintaining output and keeping their expensive machines running for the maximum possible number of hours and, particularly important, as a way of coping with fluctuations in demand. Night shifts and double-days have been worked for well over a century. Both World Wars led to a rise in shift working. The disadvantages (for workers) have always been recognized. Workers and their trade unions have always demanded premium rates of pay to compensate for the disadvantages, employers have usually accepted the need to offer a financial incentive, and workers on normal hours have considered those on shifts entitled to special recompense.

In so far as the effects of shift working were investigated systematically prior to the Second World War, the principal interest was nearly always in output and productivity. Subsequently the effects on the workers' lives and well-being became a research area. This was related to the more widespread adoption of shifts during the so-called 30 glorious years of full employment and economic growth which followed the Second World War when some firms could find markets for virtually everything that they could produce. An additional incentive was that industry was becoming more capital intensive and technological cycles were accelerating; employers were therefore keen to gain maximum output before each generation of equipment became obsolete.

The effects of shifts on leisure

Between the 1950s and 1980s the effects of shifts on workers' health, family and social lives were investigated in a series of studies (Brown, 1959; Mott *et al.*, 1965; Marsh, 1979; European Foundation, 1980; Brown and Charles, 1982; Carter and Corlett, 1982; Lloyd, 1986). The findings were crystal clear and consistent, and their predictability was probably a reason for the stream of studies drying up. Investigators noted that the effects of shifts, or unsocial hours as they came to be described, depended on the particular shift, and on the age and sex of the employee. For example, young people's lifestyles were particularly vulnerable. That said, all the studies found that shift systems were disliked: workers accepted the consequences only for the extra pay, and sometimes because the only alternative appeared to be unemployment. A common feeling was that rotating shifts were preferable to enduring permanently the inconveniences of any one unsocial schedule. Many employees accepted shifts only for a part of their working lives, when they had the greatest need to maximize their earnings.

Shifts have been found to damage employees' health. Sleep patterns are usually disrupted. Shift workers complain that they are unable to gain their normal hours of rest. Eating disorders are also common. However, shift workers have been equally vociferous in complaining about the effects on family relationships and leisure activities. Shifts de-synchronize the time schedules of family members. Individuals on shifts may be unavailable to sleep with their partners and share family meal times. They may have to miss evenings in pubs with their friends and weekend trips to sports events. Rotating shifts make it difficult for individuals to plan ahead or accept any regular social commitments: they cannot rely on being free at any particular times. Irrespective of whether their amounts of leisure time and overall levels of leisure activity are reduced, all the studies have found that shift workers complain about the lack of regularity and predictability in their daily and weekly routines, and their problems in coordinating their own time schedules with those of friends and other family members.

Exceptions

Researchers have noted some leisure advantages in working unorthodox hours. Individuals can be left with expanses of private time to pursue personal interests and hobbies when other members of their households are at work or school. However, shift workers sometimes find that such time becomes accounted for by new social obligations. For example, males can find that they are expected to care for young children during the daytime. A further advantage appreciated by some shift workers is their ability to pursue leisure activities at off-peak, less congested and sometimes cheaper times.

Shift workers in better paid, higher status jobs have been the least likely to complain about their work schedules (Roberts and Chambers, 1985). Nearly all the relevant investigations have been among manual employees. Up to now they have been more likely to work shifts than office staff. However, there are some professions where unsocial hours are customary, and it is perhaps noteworthy that these professions have not identified their work schedules as a problem requiring correction. In professions such as medicine the practitioners have probably accepted the occupational hours as part of the package. Moreover, individuals with high incomes are probably the most able to take advantage of having private time to pursue personal interests. Money can ameliorate if not solve a host of leisure problems. The relevant studies have found that the groups most vulnerable to the disadvantages of shifts are employees who are already disadvantaged such as women in low paid jobs who have to juggle the requirements of their paid occupations and domestic responsibilities (see Roberts and Chambers, 1985).

There has always been one further exception to the research findings that abnormal hours of work are experienced as inconvenient. Part-time women workers (whose working hours may not be unsocial) have consistently expressed higher levels of satisfaction with virtually all aspects of their jobs than all other groups of employees (Penn *et al.*, 1994). These women's job satisfaction appears to be related to their secondary orientations towards work; they expect less in and from their jobs than other groups of employees. In all other respects – the status of their jobs, the rates of pay, security and promotion prospects – Britain's part-time female workers (over 46% of all women in paid employment) are a notoriously disadvantaged group. Even so, they have been offered as models that the rest of the workforce may be required to emulate by spending less time in paid work, varying their amounts of paid work during the life course, and blending paid work with domestic chores and caring (Hewitt, 1993).

The Normalization of Non-standard Hours

Recent trends

One reason that research into the social effects of shifts has faltered since the 1980s is that the findings became predictable, but an equally important development has been the continued spread of non-standard working hours to the point where these have become sufficiently common to be described as normal, at least in a statistical sense. This has been due to a spread of flexibility throughout work organizations and labour markets. There are numerous signs of this trend. During the 1980s in Britain there were declines in the proportions of male and female employees in full-time permanent jobs (Brown, 1990). By the early 1990s the proportion of part-time employees had risen to approximately one-quarter of all workers and nearly half of all women employees. In 1995 Will Hutton described the 30–30–40 society, Britain, where just 40% of the working age population had permanent full-time jobs, 30% had no paid employment, and the other 30% were in jobs with non-standard terms or conditions. Meanwhile, full-time employees have been working longer than formerly. This trend is discussed in greater detail below. The point of current relevance is that hours of work are now more dispersed in terms of how many hours people work per week and their scheduling. In Britain there are more staff on part-time hours and also more working 50-plus hours per week. In the early 1990s two-fifths of all British managers were working more than 50 hours, and over half were taking work home at least once a week (Institute of Management, 1993). Self-employment has also become more common. The number of self-employed in Britain doubled from roughly 1.5 to over 3 million between the mid-1970s and the mid-1990s. By then 11% of all firms had some home-based staff. These amounted to only 2.5% of the entire workforce but their numbers are expected to rise with the spread of tele-working (Huws, 1993). Home workers and the self-employed can rarely separate work from the rest of their lives as is possible when employees have separate workplaces, and the self-employed are renowned for working as long as, and whenever is, necessary. The decline of the standard work week can be seen in that by the early 1990s less than 20% of all work establishments in Britain required no Sunday working of any kind (Bosworth, 1994), and only 34% of all employees worked 'normal' full-time hours, that is, between 8.00 am and 6.00 pm, Monday to Friday (Hewitt, 1993).

It is fair to point out that employers have always sought flexibility (Pollert, 1991). They have always wanted their staff to learn new skills when necessary, and to extend their working hours during upturns in demand, and have trimmed their workforces during recessions. However, since the 1970s firms have become even more flexible and this has applied in every European Union country (Beatson, 1995). Total hours and work patterns have become more varied. There are more people working 4 days on followed by 4 days off and on other schedules that obliterate the customary 7-day cycle. The ways in which flexibility is achieved vary from firm to firm, according to the type of business, and from country to country, but the trend has been noted everywhere. The conditions requiring flexibility are not new but have intensified and become more widespread. Businesses are increasingly capital intensive and want maximum output before their investments cease to be 'state of the art'. During recent decades additional developments have contributed to the spread of shift systems. The proportion of employment in consumer services, including leisure services, which must be produced and delivered when people want to spend and consume, has risen. The growth of leisure is among the causes of more people working odd hours. Then there is the increasingly intense, often global competition. More and more firms have reason to feel that they must be open for business whenever their customers or suppliers in Korea, or wherever, are in their offices. International competition in manufacturing, and in financial and other business services, is increasingly fierce. Technological and product cycles continue to accelerate. Firms feel that they need to be flexible in order to cope with these turbulent conditions, and this means demanding flexibility of their employees. This may mean requiring permanent core staff to be adaptable, accepting continuing education and re-training simply in order to keep abreast, and being able to change the size and shape of establishments' workforces from year to year, month to month, week to week, day to day, and sometimes from time to time during each working day. So more full-time and part-time staff are expected to work overtime, paid or unpaid, or to take work home, as and when required. Evenings and weekends have ceased to be sacrosanct leisure occasions.

In the 1960s and 1970s there was a trend towards flexitime which allows employees to select their own hours of work around a compulsory core. Such practices continue. Indeed, by the mid-1990s one in eight employees in Britain were on some type of flexitime system (Noon and Blyton, 1997). However, since the 1970s it has become more common for employers to stipulate when and for how long their staff will work. Labour market conditions, with high levels of unemployment, have enabled employers to enforce these conditions, and many have felt that they have had no alternative but to do so in order to keep abreast or ahead of competitors. So it has become more common for employees to have contracts which guarantee only part-time, if any, hours but which expect the staff to be available (on call) or to work longer, as and when required. In 1981 Stanley Parker argued that 'pure leisure' was most likely when institutions were flexible while individuals could be self-determining, but this is not how the economy has subsequently developed. Rather, it is the employing organizations that have been deciding how employees will be flexible.

Post-Fordism and the portfolio worker

These trends are sometimes seen as an aspect of the transition from a Fordist to a post-Fordist era. Under Fordism large firms and their occupational structures were among the reliabilities. Large firms survived recessions and could offer secure jobs and progressive careers. Post-Fordist flexibility changes this. Another manifestation of the new flexibility is that career ladders have been either disappearing completely or have become shorter and less secure, and fewer blue-collar and white-collar workers, skilled and other employees, are able to regard their occupations as secure for their working lives. This applies in firms of all sizes. An upshot is that working life has become discontinuous. What was once the norm for women only is now spreading throughout the workforce. Women have hitherto been at the centre of the 'revolution' in working patterns. They have been the most flexible workers. Male biographies and day-to-day ways of life have been perhaps remarkably stable alongside the changing experiences of women during the 20th century (see Hinrichs *et al.*, 1991; van der Lippe, 1996). However, men's jobs are being affected by the current changes.

One school of thought argues that the destandardization of working time, the increased variation in total hours of work, and all the other manifestations of post-Fordist flexibility, means that the future will belong to the portfolio worker. These individuals will work for many employers (though the workers may be legally self-employed) consecutively and sometimes simultaneously. No organizations will offer security and career progression. However, optimists argue that individuals will be able to find security and a sense of career development in their forever expanding portfolios of skills which ensure a demand for their labour (Bridges, 1995). Like other trends, portfolio work is likely to mean different things at different occupational levels. It is common for the same individuals to occupy several company directorships, and seats in the UK Parliament as well, in some cases. Some professionals do well as consultants. For other people, however, portfolio working is likely to mean a succession of temporary, often part-time jobs, interspersed with periods of unemployment, as individuals try to generate sufficient income to maintain their lifestyles through combinations of office cleaning, waiting-on, supermarket and similar jobs.

In the Netherlands, Koen Breedveld (1996a) has argued that other writers have exaggerated the pace and extent of change. He has analysed Netherlands time budget data in order to measure the extent to which working time has been destandardized. The proportion of all work performed outside normal hours rose from 12.7% in 1975 to 14.2% in 1995. The proportion of workers who did some work at abnormal hours in a typical week rose from 49 to 56%. Breedveld concludes that, 'Flexibilisation is definitely a hype. . . . Figures that underline this hype are all too greedily welcomed and too cheerfully reproduced by those who, for their fortune and status, depend on them' (p. 15). However, working time in the Netherlands is more regulated by law than in most European Union countries (Britain is the least regulated), and in the Netherlands these regulations were relaxed slightly only in the 1990s. Breedveld's own figures show that by the mid-1990s most employees in the Netherlands were doing some work at non-standard times.

Adjusting or tolerating the new flexibility?

Leisure research needs to catch up with these trends. There are diametrically opposed views on the likely leisure implications, but agreement that the implications will be profound. After all, the normal patterns of home life and leisure that developed after the Second World War depended on economic security and standard leisure occasions – evenings, weekends and holidays – which nearly everyone could enjoy. However, one possibility is that we are witnessing the advent of postmodern modes of time organization which, in time, will become just as acceptable as modern, industrial time.

The early factory owners had difficulty in disciplining labour to accept the new industrial work rhythms. Previously most people had lived according to so-called natural or cyclical time. Things were done when they just had to be done or could be done. This was replaced by linear, industrial time, clock time, a key feature being that there were proper times, and places as well, for most things – work, education, play and worship. The first generation of factory workers was resistant whereas by the late-19th century industrial time schedules had been accepted as simply normal.

Could we be witnessing the birth of new, postmodern patterns of time organization which, when bedded in, will be just as acceptable as linear, industrial time? Under postmodern time it is said to become acceptable to do anything at any time. So working non-customary hours could become less of a problem than in the industrial past. Television entertainment and some cinemas, shops, restaurants and bars are now open around the clock. Telephone banks never close. Cities have been developing night-time economies and becoming 24 hour places. Video-recorders enable people to record and watch any kind of entertainment at any time. In the future people may routinely bank time just as they have learnt to bank cash thereby enabling themselves to reconcile the spending and leisure patterns that they personally prefer with the ups and downs of portfolio careers. For example, by working long hours while they are economically active employees may earn sabbaticals. Average earnings and spending are now considerably higher than in the 1970s so more workers should be able to take advantage of the opportunities, and cope with the synchronization problems hitherto associated with non-standard working hours. Some couples deliberately seek different non-standard work schedules in order to maximize both the household income and the time when at least one adult will be available to care for children (Corcoran-Nantes and Roberts, 1995). This optimistic train of thought on the broader social impact of the spread of non-standard work schedules is able to regard female workers as skilled time managers from whom others must now learn (Hewitt, 1993).

The contrary and, in my opinion, more realistic viewpoint is that most people will be unable to adjust their lives to the bewildering array of work schedules and career uncertainties in which they now find themselves enmeshed, and that, rather than dissolving, the problems associated with shift work will simply become more widespread (see also Breedveld, 1994). The 7-day cycle and occasions such as the weekend have a long-standing socio-cultural appeal that may not easily be set aside. Willy Fache (1996) has deplored the spread of Sunday working in Belgium. He argues that the quality of life is impaired when there are no dependable leisure occasions, and when family members, together with work colleagues and neighbours, cease experiencing leisure at the same times. Many employees may see little point in complaining. Those coping with the problems, if this is what they are doing, are no longer a victimized minority. The problems associated with unsocial hours may have become normal without ceasing to be problems for individuals and families. Mulgan and Wilkinson (1995) have called upon the UK government to establish new time reliabilities to counterbalance the new flexibilities. They have in mind statutory rights to educational leave, parental leave and sabbaticals. If enacted, these new rights could be experienced as poor recompense for the loss of the right to work full-time in a secure job, and to enjoy evenings and weekends as family and communal leisure occasions.

The social organization of time

The trend towards greater variety in work schedules is one reason for the recent spread of interest in the social organization of time (for example, Adam, 1990). In everyday life, time is usually experienced as simply given and it is true that people cannot prevent day following night or season following season, or make any of these episodes last longer in absolute terms. However, there is nothing natural or inevitable about the division of each day into 24 hours, or these hours into minutes, let alone the separation of working and non-working time or the proportions of the day devoted to each. Time has been organized differently in different societies, at different historical periods.

Moreover, it has been argued that different modes of time organization can coexist within societies, and that this has certainly been the case in modern societies (see Lewis and Weigert, 1981). First, there have been general cultural norms which all people have been expected to recognize and respect at most times. Up to now in our society these have included occasions such as Christmas, Sunday and the surrounding weekend. Most of these general cultural norms have pre-modern origins. Working at these times has been regarded, and experienced, as abnormal. However, there have been additional modes of time organization associated with businesses, families and leisure. General cultural norms have prescribed which of these other modes of time organization should take precedence. Needing to be at work has usually been accepted as a satisfactory excuse for missing a leisure event whereas employers have been less willing to recognize leisure interests as satisfactory excuses for work absences. It has generally been regarded as selfish to place personal leisure interests ahead of family obligations. However, there have been occasions when general cultural norms have been unclear and when different sections of the population have disagreed on which time schedules should take precedence. Whether families should be expected to cope with the inconvenience of non-standard work schedules is an example.

Some writers on leisure have argued that there is, or have predicted that there will be, a trend towards leisure becoming pivotal and given precedence over other demands. We will return to these arguments in Chapter 8. Whatever the longer-term prospects, since the 1970s it has been economic restructuring, not leisure preferences, that has driven most of the changes in the organization of most people's daily and weekly lives. Family and leisure patterns have needed to adjust and the distinctions between different modes of time organization can be useful in exploring whether adjustments to what were formerly regarded as non-standard and unsocial hours of work have become less painful than in the 1970s and before, and, if so, for whom.

Another source of individualization

Greater variety in work schedules will have contributed to the individualization of leisure. Individualization is a broader social trend. As explained in Chapter 2, it arises from the national and international marketing of leisure goods and services, and also from the decline in family stability, the weakening of neighbourhood and religious communities, the decline in the large firms and industries that once dominated many local labour markets, the spread of private transport, the accelerating pace of economic and occupational change, and the more varied experiences of young and older people in post-compulsory education, part-time and full-time employment, and periods of unemployment.

In the 1950s and 1960s there was much discussion about the spread of 'privatism' – home- and family-centredness. This is not the same as individualization. Home- and family-centred lifestyles are vulnerable in the face of the kind of individualization that is now occurring. Family-centredness will be disrupted if more families have members with different work schedules. This does not mean that there will be no shared family leisure. It simply means that this will not always be as possible, or as predictable and reliable as in the past. How many 'partners' nowadays have to consult diaries to discover when they will be able to spend a full weekend together? People may not spend the rest of their leisure time alone. They may be gregarious and go out frequently, but not reliably with the same people at the same times of each week, year in and year out. Leisure companions will become temporary. Every individual will have a unique, individualized leisure career and social network, albeit based on experiences which are shared with many other people. Leisure will cease to be based on stable groups, times and places. Maybe people will be able to adjust to such situations but those who have been able to work hitherto normal hours and to enjoy normal leisure appear to have remained grateful for the opportunity.

Working Longer

Up to the 1970s working hours were in long-term decline. The reason was obvious, or so it appeared: the working population was taking some of the benefits of economic growth by enjoying more free time. This trend was expected to continue indefinitely. The 3-day weekend, and the 3-day and even the 2-day work week were forecast. The question was not so much whether as when. Automation was to reduce the need for human labour. As well as shorter work weeks people were expected to enjoy longer holidays and even sabbaticals. How wrong can you be?

The extension and intensification of work

Since the 1970s Britain's full-time employees have been lengthening their work weeks. There are more part-time employees but they are working alongside full-timers who, on average, have been extending rather than reducing their hours of work (Tyrell, 1995). This increase has been mainly among managers and professional, highly qualified employees who comprise a growing proportion of all workers (Holliday, 1996). In addition, the self-employed, among whom long hours have been customary, have grown in number. Since the 1970s professional and management staff have been working longer on average than the manual working class. It is mainly males who are spending 50 or more hours per week in their paid occupations but in the late-1980s the number of full-time female employees began to rise (Hakim, 1993). They are still very much a minority in the higher-level professional and management grades, and among the self-employed, but growing numbers of women have been joining the men in these career grades where working time has lengthened.

People in full-time jobs are working not only longer but also harder as the downsizing and delayering that began in the manual grades have spread up the occupational structure. The lunch hour appears to be a thing of the past: the average time now taken is just half an hour and more managers and professional staff spend breaks at their desks catching up with reading and paper work. Long business lunches are out. Working breakfasts have joined business dinners in extending the working day. Work has become fashionable even among the super-rich (see Seabrook, 1988). Individuals who appear to have no financial need to do so become appointed to numerous boards of directors. Having plenty of spare time, and making this visible, no longer confers status. Far more respect is earned by having a full diary into which it is difficult to fit any further engagement. The time pioneers studied by Karl Horning and his colleagues (1995) – the German employees who had voluntarily made drastic reductions in their hours of work (see Chapter 2) – have not been trendsetters in Britain (or in Germany) in recent years. The trend has not been towards people succeeding in employment then easing up. Quite the reverse: they have been working even longer. What happens to them upon retirement is discussed in Chapter 5.

The pace of life

Their longer work weeks will be part of the reason why many people feel that life has become more stressful and hectic (Holliday, 1996). People feel busier, but work pressures are not the entire explanation (see Box 3.1). People are spending more time shopping and travelling. We have more money to spend and this takes time. Between the 1960s and the 1990s in

Box 3.1. The time squeeze.

- In the 1980s real consumer spending on leisure (excluding alcohol) rose by 50%. Leisure time expanded by just 2%.
- Since the 1960s time spent shopping and related travelling has risen from an average of 40 to 70 minutes a day.
- The lunch hour is a thing of the past: the average time is now just half an hour.
- Managers' workloads have increased in recent years. Two-fifths work more than 50 hours a week. Two-thirds complain that their jobs are a source of stress.
- Between 1971 and 1991 the proportion of 7 year old children going to school unaccompanied declined from 70% to 7%.

Sources: Tyrell (1995); Hillman (1991); Institute of Management (1993).

Britain the average daily time spent shopping rose from approximately 40 to 70 minutes and the traditional gender gap virtually disappeared. Of course, shopping itself may be regarded as a leisure activity (see Chapter 7). But shopping still consumes time which ceases to be available for other purposes. Time spent on routine housework has declined and there has been some redistribution from women to men but people are spending more time on odd jobs connected with their homes, cars and gardens. Parents are spending more time supervising their children, or paying others for child care, partly because the streets are now considered dangerous (Hillman, 1991). This owes something to the increased traffic and also to the demise of neighbourhood communities in which people knew each other and felt secure. Today, the territories outside people's gardens are typically perceived as full of strangers who cannot be trusted. We are also devoting more time to personal hygiene – cleaning and adorning our bodies. How have we found the time? Firstly by sleeping less. Secondly because Britain's full-time employees, males and females, have experienced a decline in their leisure time.

International comparisons

This upturn in hours of work has not been worldwide. The trend has not occurred in all European countries. Indeed, in a European Union (EU) context Britain is the exceptional case in this respect (see Fajertag, 1996; Rosducher and Seifert, 1996). In the mid-1990s the average work week of Britain's full-time employees was three hours longer than the EU average (Noon and Blyton, 1997). The general pressure on hours of work elsewhere

in the EU in the 1990s has been downwards. This has been in a c
where organized labour, employers and governments have been unable t
identify an alternative solution to unemployment. In some countries, including Germany, employers have favoured reductions in standard working hours as an alternative to more expensive redundancies and compensation payments to workers placed on short time (see Blyton and Trinczek, 1996).

However, there is no European Union country in which this downward pressure has led to a sharp all-round decline in hours actually worked. When modest reductions have been achieved this has usually been a defensive strategy, to prevent job losses rather than to create additional jobs, and has normally been in exchange for greater flexibility (for the employers) over the scheduling of working time, plus an intensification of work. Volkswagen, where the standard work week for blue-collar and office staff was slashed to 28.8 hours in 1994, remains a very exceptional case. Its workforce accepted a 16% cut in earnings in exchange for a guarantee of no redundancies. Polls of the Volkswagen workforce found the majority expressing satisfaction with the shortened work week which created more time for leisure and their families. However, at Volkswagen it was possible to account for the greater part of the 16% drop in earnings by eliminating the former 13th month (Blyton and Trinczek, 1996).

There are some signs that continental Europe may be emulating Britain and North America. In The Netherlands there has been an overall decline in average hours of work, but Zuzanek *et al.* (1998) have shown that this is entirely due to an increase in the proportion of part-time workers. When only full-time employees are considered the same upward trend in working time noted in Britain and America becomes apparent.

The Japanese have made more recent progress than European countries in reducing their (previously comparatively long) hours of work. This process is encouraged by their government which has wanted to improve the quality of life and to stimulate domestic consumption thereby easing the country's positive trade balance which has led to so much international criticism. However, the Japanese and other Pacific Rim 'tigers' still work more hours per year than employees in other economically advanced countries. In Japan it is still considered bad form to leave work promptly at the official end of the working day. Salarymen who do not wish to raise questions about their career and company commitment would not dream of using all their holiday entitlement (Koseki, 1989; Harada, 1994). Working time in Japan has declined since the 1960s but leisure participation has risen more rapidly. Daily life has become more hectic and time budget data show that, as in Europe and North America, the Japanese have found the extra time by sleeping less (see Steger, 1996). This is the context in which public sleeping – on trains, during breaks at work, in libraries and other such places – has become acceptable conduct in Japan. Apart from enabling individuals to recuperate, public sleep enables them to signal their exhaustion and, sometimes, to create private space and imply a lack of interest in whatever else is going on.

Britain and America, the countries where the recent lengthening of hours of work has been most pronounced, appear to be emulating Japan as fast as the latter is moving towards Western norms. It may be something specific in the Anglo-American business cultures that has led to the recent lengthening of working time, especially in the management and professional grades in these countries. Other countries may or may not follow suit. In the meantime leisure researchers have been puzzled by the trends and are uncertain of the implications.

Ruthless economy or workers' preferences?

Has the ruthless economy become a threat to leisure? Juliette Schor (1991) has complained that Americans are overworked and blames, firstly, pressure from employers who have been able to make heavier demands in a business climate where staff have feared for their jobs and careers. Secondly she blames the 'addictive power of consumption'. Schor believes that leisure time is becoming a conspicuous casualty of prosperity. She has urged a search for 'new strategies', social progress beyond 'work and spend', and advocates an all-round reduction in working time which, she believes, will lead to more satisfying and environment-friendly lifestyles, and less inequality (Schor, 1998). However, before deploring the decline of leisure and the pressures that induce workers to sacrifice their leisure time, there are several points to bear in mind.

First, the fact that most workers say that they would prefer shorter hours needs to be treated with caution. In Britain less than one-third of all employees work 40 hours a week or less whereas over two-thirds say that they would like to do so. Only 7% claim to want to work in excess of 50 hours per week whereas nearly a quarter do just this. When faced with so simple a question it is hardly surprising that most people express a preference for working less. It would be no greater surprise to find that most workers would like higher pay. In practice it is often necessary to trade one value against another. Approximately two-thirds of British managers complain that their organizations confuse working long hours with commitment. Less than half say that they are happy with their current work-life balance. However, only a quarter say that they personally want to work fewer hours (Oliver, 1998). Maybe money and career advancement are now the stronger demands, and maybe this is due to the genuine attractions rather than the 'addictive power' of consumption.

Second, the same caution must be exercised in the face of evidence that long hours are stressful and, it may be inferred, bad for people's well-being. It is the case that around two-thirds of managers say that their jobs are a source of stress (Inkson and Coe, 1993). What must be borne in mind is that pressure is not necessarily damaging. It can sometimes be stimulating and challenging, even health promoting. The occupational groups with the longest hours of work score relatively well on all the standard health indicators. It is relevant here that the uses of leisure time that are most likely to relieve stress are active rather than passive (Zuzanek, 1996). Having nothing to do, when unemployed for example (see below), can be at least as corrosive as too much work. The threat of unemployment may be just as, if not more, debilitating than pressure to take on more tasks.

Third, long work weeks, relative to the current average, do not appear to depress overall levels of leisure activity (Holliday, 1996; Robinson and Godbey, 1996). Actually the occupational groups with the longest working hours have the highest participation rates across leisure activities in general. This applies when one compares the employed with the unemployed, housewives with part-time, and the latter with full-time female employees, and workers on average with those on well above average full-time schedules. The higher socio-economic strata still 'do more' at leisure even though, on average, they are also doing more work. This phenomenon has been described as 'time elasticity' (van Ophem and de Hoog, 1998).

Of course, there has to be a point at which increased working time would suppress leisure activity. If individuals worked so long that they had neither the time nor energy to do more than work, eat and sleep their leisure participation would be zero. The fact is that most people with relatively long work schedules today do not work that long, and one reason why they can have high rates of leisure participation is that the longer hours tend to be worked by the higher earners. They have the money to indulge their tastes in the leisure time that remains. Money can buy time. With sufficient cash people can do virtually whatever they want at the times that they choose. They can afford meals out which make it unnecessary to return home prior to evenings out. They can afford to have their cars repaired and serviced at garages, and to hire paid help with their homes and gardens. It is also relevant that the self-employed, managers and professionals often have considerable 'time sovereignty', that is, scope to decide exactly when they will work. They tend to work their 'odd hours' at home. When work is taken home it can usually be done at the employed or self-employed person's discretion (see Breedveld, 1996b). Being able to control one's work schedule is a huge advantage for ensuring one's availability for preferred leisure occasions and activities. Television viewing and doing nothing in particular are the uses of time that tend to be squeezed when people work relatively long hours. Participation in out-of-home recreation is unscathed and, indeed, benefits from the individuals' relatively high earnings. It is not clear that those involved are sacrificing their leisure. It is equally plausible to interpret their behaviour as maximizing their preferred leisure opportunities.

Fourth, as the population's earnings, spending power and leisure time increase, one would expect a shift in people's preferences away from yet more leisure time and towards higher incomes. Additional earnings and additional free time will both be subject to diminishing marginal utility, meaning that the more people possess, the less an extra unit of either will be valued. Hence the need to 'incentivize' senior managers and traders in financial markets with what appear, to workers in ordinary jobs, to be astronomical bonuses and pay rises. These occupations are extreme examples of a more general trend: as labour productivity and therefore earnings rise, the income sacrificed by a given reduction in working time or effort will increase. In relatively prosperous societies where there are omnipresent opportunities and invitations to consume, it is to be expected that appetites for further reductions in working hours will subside in preference for higher earnings. It is also likely that employers will become increasingly resistant to further reductions in hours of work, and keener to obtain as much input as is possible, especially from their more skilled employees who are the most expensive to train. In former times, when workers were at their jobs for 12 or more hours each working day, it is likely that trimming their hours of work had such beneficial effects on the employees' attentiveness and energy when on the job that output declined far less steeply than working time (see Chapter 2). Diminishing returns will also apply in this area. Once relatively short hours have become standard, the improvements in employee performance that accrue from further reductions are likely to be negligible.

Petolka (1996) has noted some important exceptions in Finland in firms where shift lengths were reduced to 6 hours, and the standard work week to 30 hours, in government-backed attempts to create more jobs and reduce unemployment. In six companies studied, productivity rose to a sufficient extent to justify maintaining the employees' previous earnings. However, this was in a context where the workers feared both unemployment and significant reductions in their normal earnings. Also, there had obviously been sufficient slack in the earlier work regimes for the employees to speed up when given an incentive. Given the work intensification that has occurred in most Western companies in recent years, there must be many workers who would feel unable to raise their productivity as a way of paying for shorter working hours. This will mean that someone, the employer or the employee, will have to sacrifice income pro rata. The more highly paid the employees, the more someone will stand to lose. Employers who have invested heavily in training skilled labour are likely to treat such employees like expensive machines and want to keep them working for as long as is consistent with the required quality of work. If the employees have to pay for their own education and training this will intensify their desire, and need, to maximize their earnings in order to obtain returns that justify their investments in human capital.

Fifth, there are some jobs, generally higher-level management and professional positions, in which, provided all other things remain equal, practitioners become more knowledgeable and effective by reading as much as they can, and discussing ideas with as many colleagues as possible. Senior managers are best able to run their organizations if they have frequent contact and communication with colleagues and subordinates. Politicians can stay on top of their jobs only by talking constantly to each other, and being available to constituents. It is not necessarily workaholism; it is just as likely to be rational assessments of what the jobs require, and the cost-benefits of increases and reductions in working time, that keep the people concerned working for over 50 hours a week. Given all this, the 'time pioneers' studied by Horning and his colleagues (1995) in Germany seem more likely to remain deviants than to become trendsetters.

An uneven trend

A final set of points to consider stems from the lengthening of working time affecting only some occupational groups, and some of the workers within the groups in question. Leisure time is not contracting throughout the entire population (Gershuny, 1986, 1992). There is no general mounting leisure time famine. Gershuny has argued that people's ability to choose and control their hours of work will be sufficient to check any trends towards unwanted feasts or famines of leisure time. Across Europe the total amounts of time spent in paid and unpaid work have been fairly stable in recent decades (Mogenson, 1990). The main general change has been a narrowing of gender differences in time spent in both paid and unpaid work. Moreover, the recent rise in hours worked by some occupational groups in some countries is not the radical break with former trends that it might at first appear. As noted in Chapter 2, in the early 1960s Wilenski (1963) observed that most of the time released from paid work during the previous century had been used to enable some groups (mainly the young and the old) to withdraw from the workforce altogether rather than to reduce the working time of those who remained in employment. And the decline in hours worked up to the 1970s occurred mainly in the manual grades, who were then the majority of all employees. In many professions there has been no decline in working time since the 19th century. Staffen Linder's seminal book, *The Harried Leisure Class*, was first published in 1970. The main reason why life is becoming more hectic for many people is not that they are working longer but because their leisure time is increasing more slowly than their spending power and the things that they want, and can afford to do. Between 1979 and 1989 in Britain leisure spending (excluding alcohol) rose by 50% in real terms while leisure time grew by only 2% (Tyrell, 1995).

The recent lengthening of working time is probably less of a threat to leisure than the former's destandardization, especially when employees have little discretion over exactly when they will work and for how many hours, if they have jobs at all. Under these circumstances the destandardization of working time is likely to play havoc with people's preferred leisure schedules. Moreover, job insecurity has many of the same disastrous consequences for personal well-being as unemployment (see below and also Gallie *et al.*, 1994).

The current trends in working time are certainly helping to create a more differentiated society. The growth of leisure has ceased to mean marginal gains in free time and spending power, and more opportunities to participate in leisure activities, for all sections of the population. Some people are working longer and harder and have less leisure time than formerly but have the incomes that enable them to be highly active in the leisure time that they retain. Their leisure may adjust rather than suffer if more shops, banks, cinemas, restaurants, government offices and so on can be persuaded to stay open for business 24 hours a day, 7 days a week. However, members of this section of the population are likely to suffer from a shortage of genuine 'spare time', private personal time when they can respond positively to any opportunities or invitations that arise either in their work or leisure. Literally free time is a likely casualty of trends among the hard working high earners.

Other sections of the population are working unsocial, though not necessarily long hours in the generally low paid occupations that supply the leisure goods and services sought by the affluent. Hours of work do not need to be long in order to be oppressive. Studies of manual workers have consistently found that work typically dominates their lives: doing it, preparing for it and recovering from it. Extremely private, home-centred lifestyles have been more likely to arise from a lack of money, time or energy to go out than from free choice (Pearson, 1977; Devine, 1992). Even though their working hours are now shorter on average, the routines of manual employees are probably less compatible with satisfying leisure than the longer work weeks of managers and the self-employed. This is especially so far those in families where husbands and wives both work in arduous jobs, either or both on shifts, and where most of the household income is accounted for by rent or mortgage repayments and other routine expenses.

A third section of the population has no paid work. These individuals may have plenty of spare time but are usually poor and have lower rates of participation than all other sections of the population in virtually all forms of out-of-home recreation that cost money.

Unemployment

The return of unemployment

All capitalist economies have unemployment, but the 30 years that followed 1945 are commonly referred to as an era of full employment. This is because, outside particularly depressed regions and especially vulnerable groups such as the severely disabled, unemployment was usually short term. Some individuals left school or jobs and had spells of unemployment before they gained or regained work but these jobless episodes seemed unlikely to make a lasting impression on the individuals' minds or lifestyles. In contrast, since the 1970s levels of unemployment have risen in virtually all modern societies and, as this has happened, long-term and repeated unemployment have become increasingly common. This does not only affect the people without jobs; as mentioned earlier, feelings of insecurity spread to those in employment (see Gallie *et al.*, 1994). In the study of leisure, interest in the implications of having no job has partly replaced the former interest in the leisure effects of particular kinds of employment.

The reasons why unemployment has risen are strictly beyond the scope of this book, and likewise the reasons why certain groups are particularly vulnerable, though it is relevant to note that unemployment, and therefore whatever the leisure effects might be, is highest among the age groups at each end of normal working life, and among the least qualified and skilled. The wider academic and policy making communities have displayed considerable interest in what leisure research may reveal about the social and psychological consequences of unemployment, and whether these could be changed by leisure activities and provisions. These are issues on which leisure researchers can command substantial audiences. Leisure research certainly has much to contribute to debates on whether persistent unemployment is creating an underclass or swelling the size of excluded groups. If such a stratum or strata exist they will be distinguished not just by their unemployment and relative poverty but also by distinctive cultures; behaviour patterns and attitudes which in themselves would impede the members' integration into the workforce even if other obstacles were removed. Leisure research can make important contributions to all debates about the reproduction of old and the formation of new social classes at all levels in the social hierarchy.

Questions for leisure research

In studying unemployment, leisure research has addressed two main questions. The first concerns those leisure effects where it quickly becomes apparent that it is necessary to distinguish between the effects of unemployment and the effects of non-employment. The unemployed are individuals who want work, are seeking it, and expect to be, and are expected by others to be, in employment, but who are unable to find jobs that they are able and wish to enter. Their situations and problems are different from those of other non-employed groups such as children, students, housewives and the retired. The second question is whether leisure activities and provisions can solve or ameliorate any of the problems that otherwise arise from unemployment.

Both of these questions have been answered unequivocally. All the research evidence points clearly to the same answers. Unemployment is bad for people's leisure. In fact unemployment seems to be disastrous for most things that people value – their physical and mental health, and their family lives, for example. Divorce and separation rates are twice as high in households where someone is unemployed as throughout the population in general.

The functions of employment

A good way of grasping the implications of unemployment is to consider the functions that jobs normally perform for people who are in work. First, employment provides income. This is both obvious and extremely important. On average the incomes of the unemployed in the UK are less than 60% of their normal earnings. A minority are little better off when they are in employment but these are the working poor rather than people who live well on welfare. Redundancy payments and savings may cushion the impact of short-term unemployment but if unemployment drags on savings are exhausted and households have to rely entirely on social security. This usually means doing without things that are taken for granted by the working population such as buying Christmas and birthday presents, being able to share rounds of drinks, and coping with the incidental expenses that are involved in attending most meetings of clubs and societies. Long-term unemployment invariably means long-term poverty.

A second function of employment is to supply what psychologists call 'categories of experience' which are good for mental well-being (see Jahoda, 1982). Jobs give people something to do. Irrespective of whether people claim to like their jobs, research suggests that almost any activity is preferable to idleness. Going to work also supplies social contact which, once again, is conducive to well-being. Individuals' occupations also set goals and enable them to experience a sense of individual and group achievement. Of course, jobs differ in the degree to which these benefits are conferred, but all the research evidence suggests that almost any occupation is better than none.

Third, employment creates a structure for workers' lives. It imposes a pattern on the day, week, year and lifetime. Emile Durkheim (1858–1917) was the founding father in sociology who recognized that individuals need structure and that total freedom is not a recipe for happiness. Durkheim argued that, in structureless situations, individuals experience anomie, a state of normlessness. They feel restless and an indication of their distress is a rise in the rate of suicide. A common complaint by the unemployed is that their time 'hangs'. Their lives can be so bereft of structure that signing on at the employment exchange may be experienced as a major highlight. It is necessary to have a job to experience the Friday night feeling. Without a work week the weekend becomes meaningless. Without working weeks there can be no holidays.

A fourth function of employment is to supply status and identity. Needless to say, occupations differ in status, but any job gives an individual a position in society. We place people by their occupations. Age, sex and marital status are other major social markers, but the everyday getting acquainted question, 'What do you do?' is ordinarily taken to mean, 'What work?' The public at large may agree when questioned in surveys that the level of unemployment is due to economic conditions and, perhaps, government policies and that the unemployed should not be blamed. Nevertheless, in everyday social interaction the unemployed are made to feel responsible for their situations. Other people have jobs despite the economic situation and government. Encounters with acquaintances usually lead to inquiries as to whether a person has found work yet. Having to admit failure is painful. The unemployed complain that their self-respect is undermined, that they are made to feel non-persons and that they do not count. To avoid these injuries the unemployed have been known to conceal their predicaments from neighbours and even members of their own families by continuing to leave home every morning as if they had jobs to go to. For many people their occupations are not just things that they do but also who they feel they are: teachers, local government officers, salespersons, secretaries or whatever. Our occupations become part of our self-concepts and their loss can remove one of the props that keep our personalities intact. The many functions that employment performs mean that unemployment usually inflicts psychological and social as well as economic damage.

Simply thinking carefully about the nature of leisure on the one hand, and the functions of employment on the other, suggests that leisure activities will not be satisfactory substitutes for jobs. Leisure activities, if the unemployed participate, may provide some valuable categories of experience – keeping busy, social contact and achieving goals. However, leisure activities are usually no substitutes, or extremely poor substitutes, in respect of all the other functions of employment. First, leisure activities are not income generating. It may be possible for some of the unemployed to turn hobbies – gardening, sport or music – into paid occupations, but in such cases the activities become work rather than leisure. Leisure activities cannot normally structure time in the same way as paid occupations. All leisure activities are voluntary, or at least not obligatory: it is not absolutely necessary to do them. Some individuals may use leisure activities to give their lives a structure but this requires considerable self-discipline whereas paid jobs 'have' to be done. For some individuals leisure activities may be developed into satisfactory bases for social and personal identities. Some international sports players use unemployment benefit in lieu of sports scholarships. Outstanding performers can use leisure activities to create socially esteemed roles and identities for themselves. But this is possible only for a minority. Leisure 'work' is usually valued only by members of the relevant interest groups. Becoming a competent golfer does not command the same general social recognition as having a paid occupation. Pay is important for the standard of living that it permits and also for signifying that one's contribution is valued by the wider society.

Unemployment careers

The full deprivations of unemployment do not take effect the instant a jobless episode commences. Stages in unemployed careers have been identified. In the early days and weeks of being unemployed individuals' morale can remain high. People typically retain their normal time structures and regard themselves as 'looking for work' rather than 'out of work'. They may have savings that allow normal patterns of spending to be preserved. They often keep themselves busy by doing outstanding jobs around their homes. For a time the unemployed may make an occupation out of job searching.

Many escape from unemployment quickly, but those who fail to do so tend to become frustrated. They become bitter when their job applications receive only formal acknowledgements, if that. Once homes have been redecorated and gardens tidied up, and as their savings are exhausted, people begin to question what they had formerly taken for granted. They are likely to question their job search tactics, the advice that they may have been given, whether they really have any future chances of employment, and eventually they are likely to question their own worth.

At a later point, if their unemployment continues, they are likely to become resigned to the situation. Their job searching then becomes nominal. They become afraid to apply for jobs because repeated rejections are painful. The poverty and the stigma of unemployment inhibit social interaction. Daily routines geared to employment collapse. Individuals cease trying to keep up appearances. They become broken economically, socially and psychologically.

The leisure of the unemployed

Rather than compensating for their lack of work by ploughing their time and energies into leisure activities, the unemployed tend to reduce their levels of leisure participation. Some have responded to leisure researchers' questions by replying, 'What leisure?' (Hendry *et al.*, 1984). Many find the concept completely inapplicable to their own lives.

The unemployed may protect their leisure by maintaining the range of their activities while reducing their frequency of participation, or they may maintain the frequency while narrowing the range, or they may do things more cheaply than when they were wage earning. Holidays away, nights out and new outfits can be purchased at relatively low cost. However, all the studies show that in some way or another unemployment depresses leisure participation (Stokes, 1983; Kelvin *et al.*, 1984; Raymond, 1984;

Roberts *et al.*, 1987; Kelly and Raymond, 1988; Roberts *et al.*, 1991; Lobo, 1993). This is through a combination of financial constraints, the tendency for the unemployed's social networks to narrow as they lose contact with working friends, and their lack of status which can make any social exposure threatening and distressing. Inevitably, it is leisure that costs money which tends to be trimmed or eliminated. Compared with the working population's leisure, the unemployed's is more home based. They spend more time looking for work, doing household chores, in bed, reading, resting, watching television, listening to the radio and gardening. Television and other mass media feature prominently, though not necessarily as sources of immense satisfaction, in the unemployed's daily routines. Going out is inhibited by their inability to afford the costs of transport and incidental expenses associated even with nominally free out-of-home activities. This is how the unemployed lose contact with former friends, and sometimes make excuses for non-attendance at family celebrations due to their lack of appropriate clothing and the other indignities connected with exposing their impoverished unemployed selves.

Leisure provisions for the unemployed

In virtually all parts of the UK, and in all other Western countries, there have been leisure policies and programmes targeted at the unemployed. These can take the form of special sessions, reduced charges or even free admission (usually at off-peak times) to sports centres and other facilities. The impact of these measures has been studied systematically and two conclusions can be drawn.

First, the programmes do not reach the majority of the unemployed, and most of those who attend do not become regulars (Glyptis, 1989). The unemployed are notoriously difficult to mobilize whether for political or recreational purposes. This is partly because the unemployed are a stream rather than a static group. Individuals who have recently joined the stream may not wish to identify, or to be identified, with 'the unemployed'. Their aims and efforts are geared to seeking personal escapes rather than joining others who share their unemployment problem. Those who fail to escape tend to become resigned, apathetic and fatalistic. The difficulties in tempting the unemployed into recreational activity are compounded by the fact that unemployment is highest among the least qualified and least skilled strata where normal levels of recreational activity are relatively low (see below). Individuals are more likely to continue to practise an activity during unemployment than to try something new, and the unemployed typically enter this situation with modest reserves of 'leisure capital' (Roberts and Brodie, 1992).

The second conclusion from studies of the unemployed's leisure participation, within and outside special programmes, is that high levels of activity, when maintained or generated, improve individuals' well-being but not to the level normal among the employed (Roberts *et al.*, 1982; Kilpatrick and Trew, 1985; Kay, 1987; Roberts *et al.*, 1989; Evans and Haworth, 1991; Haworth and Drucker, 1991; Haworth, 1993). Participation in recreation improves the unemployed's leisure, and this is appreciated, but it is rarely considered an adequate substitute for employment. The clear message from research is that the unemployed want jobs and wages, not leisure opportunities. It is possible to dismiss leisure programmes for the unemployed as 'bread and circuses'. The criticism has some foundation. Leisure, it appears, can be a palliative but not a solution to the unemployed's main problems. Leisure programmes, in so far as they succeed, may 'keep the lid on' without addressing the root sources of the unemployed's discontent. Successful programmes may suppress rage and adjust the unemployed to their situations thereby reinforcing their social separation and exclusion.

Fear of rage certainly seems to have been a factor in public spending on leisure facilities. The 1981 riots were followed by new sports centres in Brixton and Toxteth, and it is difficult to believe that Belfast would be the UK's, and possibly Europe's, best provided city in terms of indoor sports facilities had there been no 'troubles' (see Roberts *et al.*, 1989a).

An underclass?

The leisure evidence does not suggest that the unemployed, or even just the long-term unemployed, are becoming an underclass in the senses defined earlier, namely, possessing a different way of life and values which perpetuate their separation from the working population. This is not to say that minorities within these larger aggregates might not constitute an underclass. However, the leisure evidence does not indicate a general tendency for the unemployed to adopt values, attitudes and behaviour patterns which not only set and keep them apart, but also make them want to remain apart from mainstream society. The leisure of the unemployed is impoverished rather than qualitatively different. There are no signs in the leisure evidence of even a substantial minority of the longer-term unemployed having alternative sources of income or developing a preference for lifestyles that set them outside the wider society's legal and moral frameworks. When given the opportunity to become employed, most of the unemployed respond enthusiastically (Marsden, 1982). This is hardly surprising. The account of the unemployed's leisure presented above is unlikely to have persuaded anyone that unemployment is an attractive condition. It is true that some of the young unemployed 'work for a bit in order to do nothing for a while' and regard this way of life as preferable to continuous 'slave labour' (see Roberts *et al.*, 1982). However, the first choice of most such individuals would be good jobs and their next-best way of life is based on sub-employment, working intermittently, not long-term unemployment. Their way of life is only viable in conditions of near full employment when the individuals know that they can return to work if they seek jobs actively and lower their standards as regards the types of employment they are prepared to accept.

A future leisure solution?

Up to now leisure may not have been a solution to the unemployed's problems. Even so, some argue that leisure could become a solution. Some insist that leisure must become a solution. At any rate, they demand that we continue to search for ways in which leisure might solve the unemployed's problems since all other proposed solutions look at least equally frail.

Those who believe that leisure could or might be a solution to unemployment point out that there are exceptions to the general rule that joblessness damages people's leisure. The same applies with health. Some people's health improves when they escape from injurious work environments. In general work may be good for people's well-being but there are exceptions, when jobs are extremely stressful or demeaning for example. Tony Walter (1985) has presented himself as an exception. He used his unemployment to complete an academic thesis and claims to have been content with his circumstances. Furthermore, he claims that he was part of a significant minority who are surrounded by a conspiracy of silence. Walters accuses politicians of all conventional ideological persuasions of finding it impossible to concede that unemployment might be beneficial for some people. He also accuses researchers of ignoring the exceptions instead of examining these cases in detail and discovering how they might be made more common.

Fryer and Payne (1984) have been exceptions to this rule. They identified and studied a small sample of 11 exceptional individuals in Sheffield who had approached unemployment proactively, meaning that they had planned ahead and taken charge of their lives instead of just reacting to events. Most of the individuals whom Fryer and Payne studied appeared to be putting their unemployment to good use by taking educational courses, training for new careers, or undertaking voluntary community work for example. Could such currently exceptional cases become more common? It is necessary to bear in mind that there are exceptions to all generalizations in social science, and it is usually not the case that there is a simple and practical formula for transforming the exceptions into the rule. It is perhaps noteworthy that Fryer and Payne admit that it was very difficult to locate their 11 cases. It is perhaps equally pertinent to the present discussion that most of the individuals who had found ways of coping during unemployment were involving themselves in other forms of work rather than leisure activities.

It is sometimes argued that people would find leisure a satisfying alternative to employment if only they would abandon the work ethic and adopt leisure values. 'If only' really makes such an exercise sound far too simple. We have seen that it can be their leisure interests that motivate people to work and earn as much as possible. Furthermore, in a study in Israel, Boas Shamir (1985) found that the work ethic tended to be an asset rather than a handicap in coping with unemployment: it kept people busy, searching for jobs among other things, and therefore assisted both their psychological survival during unemployment and eventual escapes.

Clive Jenkins and Barrie Sherman (1981) have offered one of the more plausible leisure solutions. They believe that we are facing a long-term collapse of work due to the latest technological revolution and that the result will be a 'leisure shock' unless we devise an attractive leisure alternative. The Jenkins and Sherman alternative would involve a greater emphasis on leisure interests and activities in education, and increased public investment in high quality leisure services which would be cheap to users. Under these conditions they believe that individuals would seek to reduce their hours of work in order to benefit from the new opportunities to express their leisure interests, and that paid work could therefore be spread around a larger number of employees thereby soaking up unemployment.

One might query the political realism of these ideas given the recent willingness of the more highly qualified, skilled and better paid sections of the UK and North American workforces to extend their working time. However, a lifetime perspective on hours of work opens up a wider range of ways in which working time might be reduced – via education and training sabbaticals, parental leave and early retirement, for example – as well as by trimming the standard work week (Hoffman, 1996). And we have already seen that in most European Union countries in the 1990s the overt pressure on working time has been downwards. Britain is the European Union's odd case. It is exceptional, American rather than European, in the extent to which bargaining between trade unions and employers has shifted from national to firm and plant levels, and (up to 1998) in the absence of legally fixed floors and ceilings on levels of pay and hours of work. The UK also has a distinctive vision of Europe as a common market within which, as in other world markets, the separate sovereign states compete for shares. Other European Union countries recoil at the prospect of such competition driving down wage levels, intensifying workloads, and driving up hours of work. They are not attracted by the British model which produces the widest income inequalities, and the longest work weeks in the European Union. Britain's reply has been that its own is the only course that will enable Europe to compete with the rest of the world, that acceptable hours of work and rates of pay are best judged by separate employers and employees given their market situations, and that even poor quality jobs are preferable to no jobs, not least in the former being able to act as stepping stones to something better. These arguments are likely to run for many more years.

Even in Britain and North America, despite their current implausible appearances, leisure solutions to unemployment look no more difficult than proposals to solve the problem through economic growth and job generation. Maybe leisure solutions will not work for all those currently unemployed but could operate satisfactorily among specific sections of the population. For example, Robert Stebbins (1992) has argued that people with serious leisure interests (see Chapter 6) are likely to be able to cope with, and may welcome temporary or even permanent respites from paid work. Some sections of the population appear better able than others to cope with joblessness. Some individuals take early retirement voluntarily when given the opportunity (see Chapter 5). Women with young children are less likely to suffer socio-psychological damage during unemployment than most other sections of the population since they have alternative roles as housewives and mothers which structure their time, confer a social position and acceptable identities. Young people appear to suffer less damage from unemployment than older age groups. School-leavers have no established occupational identities to shatter, they can be optimistic for their long-term prospects, their families often cushion the financial deprivations, and alternative statuses are available in education and training. Leisure solutions, and other solutions for that matter, will not need to work with everyone in order to contribute to the elimination of involuntary, painful unemployment.

Eastern Europe

Like other conclusions in this and other chapters, the conclusions about the damaging effects of unemployment, and the inability of leisure to offer solutions, are likely to be specific to Western countries with their particular kinds of work and leisure. There is evidence that the impact of unemployment has been rather different in East-Central Europe in the immediate aftermath of the collapse of communism. A 1993 study of young adults in three regions of Poland found little difference between the leisure possessions and activities (except holidays) of the unemployed and individuals in jobs (see Table 3.1). It was possible to compare the findings from this Polish research with similar evidence from studies of young people in Britain in the late 1980s. Compared with British patterns, the leisure of the young adults in Poland was less commercial and more family-centred and communal. This seemed to be why unemployment in Poland was not exerting the negative leisure effects that have been recorded in all the relevant studies in Western countries. This is an example of the findings from Western leisure research being specific to Western countries. However, at the time of the Polish research the country's economy and its people's ways of life were experiencing rapid commercialization. The likelihood was that before long the effects of unemployment would become Western.

Table 3.1. Unemployment and the maintenance of normal leisure in Poland, 1993.

	22–24 Year olds (%)	
	Employed	Unemployed
Have use of:		
Car	48	40
Video	61	56
Record/cassette/disc player	90	93
Satellite dish	26	24
Take part at least once a week:		
Youth clubs or group	10	11
Play Sport	27	27
Pubs, cafes	16	15
Cinema	—	1
Watch sport	3	2
Church	26	26
Last 12 months:		
Holiday	59	46

Source: Roberts and Jung (1995).

Money, Leisure and Social Class

Lack of income is one deprivation of unemployment. It is one reason, though it is not the sole reason, why the unemployed's leisure tends to be impoverished. This is an example of an effect being over-determined; several sufficient causes are operative. Even if unemployment created no additional barriers, the drop in income that usually follows would require the victims to cut back, at least on those forms of leisure that cost money.

Income effects

Income inequalities are an important direct source, and an easily understood source, of the leisure inequalities between the unemployed and the employed, and within the working population. As noted at the beginning of this chapter, income is a major, long-established and persistent source of work effects on leisure. Occupations differ from one another in numerous ways, but some of the most glaring inequalities are in remuneration. Pay has always been the most plausible explanation of why the higher occupational strata do more. Income is the source of the clearest, widest and most consistent leisure contrasts throughout the population. Income effects are so powerful that they override what, all other things being equal, would presumably be the negative leisure effects of relatively long hours of work.

Moving from the base to the top of the socio-economic hierarchy, levels of leisure participation rise progressively. The better paid groups take part in more activities, participate in them the most frequently and/or spend the most money on each occasion. This applies in sport, the arts and tourism – all the main types of leisure activity that normally cost money. The main exception, and the source of the extra time that the better paid need in order to lead their busy leisure lives, is television viewing. The long arm of the job exerts some of its most powerful leisure effects via pay. These effects are long-standing and show no sign of diminishing. Changes in the nature of work and other social trends may have blurred some former work–leisure links, but this is not the case with income inequalities. Poor people's leisure tends to be uniform. It is dominated by low cost, time consuming, home-based activities, most notably watching television. The better-off are able to do more of everything that costs money, and there is much more variety in their leisure. They are able to build on their particular interests, often related to age and gender, in ways that are simply impossible when people are poor (see van Ophem and de Hoog, 1998).

Class-based and class-related leisure

It is more accurate to speak of present-day leisure as class related than class based. By far the strongest class relationship nowadays is that the higher strata do more of virtually everything that has a cash price. It has become more difficult than in the past to identify qualitatively distinct leisure patterns that are typical of entire social classes or even specific occupations. The main differences are now quantitative and are maintained primarily through financial inequalities.

This has not always been the case. As explained in Chapter 2, historians have stressed the strength and persistence of qualitatively distinct class leisure patterns throughout the 19th century and well into the 20th century (Howkins and Lowerson, 1979; Davies, 1992; Mason, 1994). In Victorian Britain the middle and working classes lived in different districts, dressed differently, and spent their leisure in entirely different places, and this continued up to the Second World War. There were changes in the leisure of all social classes related to rising standards of living, shorter working hours (for manual workers) and new leisure opportunities (flying, driving, dancing, holiday making, radio and the cinema for example) but without breaking down the divisions between the classes.

According to Victoria de Grazia (1992) these social class differences persisted in the rural parts of Western Europe until the 1950s. Until then leisure in Europe's rural villages was typically communal. Different classes had their separate ways of life which came together, but were not fused, on days of national and religious celebration. The leisure ideal that the authorities (fascist, communist and liberal) promoted was of loyal citizens, members of different classes, each of which contributed to the strength of the society. Victoria de Grazia explains how, from the 1950s onwards, communal forms of leisure began to be undermined as affluence and consumerism spread from the industrial classes and urban areas into Western Europe's countryside. She argues that leisure has subsequently become a sphere of choice for individuals and households who spend their time and money in a variety of very different ways. In the Italian village whose history de Grazia traces in detail, the old communal festivities were being preserved primarily as tourist attractions.

Privileged omnivores

In overwhelmingly urban Britain the social and economic trends described earlier in this chapter will have diluted spillover and compensatory work–leisure relationships. Trends towards home-centred lifestyles, prosperity that has given some households a wider range of options, unemployment which has impoverished others, and the normalization of non-standard work schedules will have conspired to undermine ways of working and playing that once characterized entire communities and even social classes. Of course, there are still pockets where traditional working class communities survive. Sections of the super rich maintain exclusive uses of leisure. The development of distinctive lifestyles has been noted among sections of the new upper middle classes, and these developments are discussed in Chapter 8. But the main class difference today is simply that economically advantaged groups have the wider opportunities and do more.

Mass affluence has reduced the number of leisure activities that are exclusive to privileged elites. Meanwhile, elite groups have been embracing popular culture. There was a time when the economically privileged tended to have distinctly highbrow tastes. They could be regarded as snobs in so far as enjoyment of the popular arts could lead to a loss of status within the groups in question. Those days have gone, partly as a result of the volume of upward mobility: the number of people who now rise into the higher socio-economic strata, typically bringing low-brow tastes with them (see van Eijek and van Rees, 1998). Today the economically privileged strata tend to be 'cultural omnivores' (see Peterson and Kern, 1996). It is true that they purchase much more classical music than the working class, but the economically privileged spend more on popular music than classics (see Longhurst, 1996). In fact they spend more on popular music than the working class. The higher strata are not distinguished by their exclusive tastes so much as by their sheer variety (see Erickson, 1996). The working class is hardly represented in the audience for high culture, but this is not because it has its own equally rich alternatives. The less well-off simply have the narrower range of tastes and activities. Money is not the sole reason. The upper strata's breadth of interests owes something to their education and their wide-ranging social networks. However, expressing their tastes involves spending money, and a reason why poorer sections of the population cannot widen their tastes and interests to a similar extent is that they cannot afford to do so.

An outcome of all this is a leisure democracy in the sense that members of all social strata do similar things in their leisure, but democracy is not the same as equality: the privileged classes are distinguished by their ability to do more which they exercise in virtually all areas of out-of-home leisure. The unemployed are an extreme case of leisure disadvantage rather than a stratum that has developed, or is likely to develop a qualitatively distinct way of life. Among the unemployed, and at all other occupational levels, uses of leisure no longer integrate most individuals into culturally distinct social classes which may then either celebrate or challenge the wider socio-economic-political order. Leisure, and many other areas of social experience, have been individualised. Money is now at the root of the main differences between uses of leisure in different social strata, and the leisure differences between them are basically and blatantly inequalities rather than alternative ways of life.

Chapter 4

Gender

Critiques of the 'Malestream'

The neglect of gender and women

This is a topic where any overview risks becoming outdated between writing and publication. Gender and leisure is a fast moving, high output research area. Things were not always so. Until the 1980s gender differences in general, and women's lives in particular, were neglected in leisure research. Textbooks dealt with these topics briefly, often as a sub-area within leisure and the family. In 1977 Liz Stanley's review of leisure research abstracts found that only 3% referred to sex differences. Why was gender neglected? It was simply not an issue for most researchers. Leisure research shared a tacit assumption with most other social sciences that women had already been emancipated and the remaining differences in males' and females' lives were either natural or socially inevitable. Until the 1970s there were few dissenting voices. Needless to say, it is very likely that a principal reason was that most researchers were men.

Normalizing the masculine

By the mid-1980s a formidable critique of conventional 'malestream' leisure research had been mounted. The neglect of women was the starting point, but not the sole telling criticism. Large-scale surveys of the public's leisure had always gathered information from and about both women and men. The bias had arisen when researchers studied trends in working time and leisure (see Chapter 2), the leisure effects of different kinds of employment, or not having any job, or when they explored youth cultures, for example. In all these areas the focus was far more likely to have been on males than females. There were many more studies of young men's than young women's leisure. Part of the explanation was that males were considered the main problem – the main perpetrators of thefts, vandalism and other anti-social acts. It was also relatively easy for fieldworkers to establish contact with male peer groups on the streets. Females were more likely to be at home. They were less visible. The public spheres of work, politics and leisure were all dominated by men. Their lives obviously appeared the more interesting, judging by the research attention that they attracted. Women were often glimpsed in research reports only through the eyes and lives of the principal male subjects. Males and females received equal attention only in studies of leisure and the family which tended to reinforce the impression that the family was a woman's proper place.

However, what the critics, who became increasingly numerous from the 1970s onwards, really resented was that research findings from male subjects were typically presented as if they were gender neutral. Males were treated as straightforward normal people rather than specifically masculine creatures. The leisure (and other) effects of particular types of education, employment, levels of income, unemployment and so on were rarely interpreted as being mediated by the subjects' masculinity. Male behaviour and responses to particular situations were treated as just normal. So when women were studied, any peculiarities in their behaviour seemed to require a special gender explanation.

A male concept of leisure?

It was also argued that researchers were operating with a male concept of leisure which was difficult to map onto women's lives. The residual concept, the definition adopted in this book, which roughly equates leisure with time that is left over when other things have been done, may more or less correspond with the realities of life for most men. The concept looks far less reasonable when confronted by women with child care and other domestic responsibilities. Do they have any genuinely free time? The titles of the books which ask questions such as *What leisure?* (Green *et al.*, 1990) and *All work and no play?* (Deem, 1986) arise from the reactions of women who have been unable to recognize their researchers' concept of leisure in their own lives.

It is more difficult with women than with men to measure their amounts of leisure time. Researchers have now shown that women's leisure activities and experiences are often entwined with or derived from paid and unpaid work. There are similar problems with trying to measure women's leisure participation with the standard checklists of activities. When males have visited sports centres and cinemas it may be reasonable to assume that they will have gone to pursue their own leisure interests. Women are more likely to have been accompanying other family members. Holidays are not exactly the same experience for women and men (see Deem, 1996). Research has now shown that women's own pleasures are often extracted from sharing the activities and experiences of male partners and children. Critics of conventional approaches in leisure research have urged the adoption of concepts and methods more in keeping with the realities of women's lives. As explained below, there has been progress, but as yet no wholly satisfactory solutions to the entire list of criticisms.

The feminist response

This is despite the fact that in the 15 years up to the mid-1990s gender differences, and women's situations in particular, developed as the leading area in leisure research. There is now a massive and still fast growing literature. Much of the impetus, needless to say, has arisen outside leisure research itself. The impetus has been supplied by the women's movements that have made gender a live, and a lively, issue not only in the study of leisure but also in research into education, labour markets, politics and so on. The leading contributors in the research and debates on leisure and gender have been women, most of whom, I think, will accept their description as (various types of) feminists. Many male researchers have been envious: they would love to be able to write and campaign for, and identify closely with, the interests of half of their fellow citizens.

Today it is impossible to claim that gender differences or women are neglected in the study of leisure. Rather, they are now swamped by a level of attention which makes it increasingly difficult to be original. That said, this has been a field in which research has not merely collected more and more facts but where there has been clear conceptual and theoretical progress in response to the original criticisms, and also to perceived inadequacies in the initial responses. The issues have been reconceptualized repeatedly as research into successive sets of questions has led to the questions themselves being re-appraised or enlarged.

It is possible to distinguish a series of streams that have flowed during the successive waves of research into leisure and gender. Indeed, it seems necessary to identify several streams in order to organize the current state of knowledge. This is despite the fact that the separation of streams is fraught with dangers. First, the most recent is unlikely to hold the status for long in such a fast moving field. Second, it is impossible to link each stream to precise dates since each set of issues has arisen gradually from, and has joined rather than replaced, its predecessors which have continued to flow in parallel. Third, it can be misleading to attach names to the streams, or even to stages in their development, since the longer-term researchers have developed their own positions and shared the general progress of their field of enquiry.

A prominent theme in the early studies, those conducted up to the mid-1980s that were able to argue vehemently against the previous neglect of women and gender in leisure research, was that women were the disadvantaged sex. Women were portrayed as being heavily disadvantaged in their leisure, and these disadvantages were often seen as just one set of manifestations of patriarchy (male power) which had pervaded all historical and existing societies. The leading issues in this research were the ways in which women's leisure was disadvantaged (by the availability of time, and the range and frequency of their activities for example), the sources of these disadvantages, and the kinds of leisure policies and provisions that would promote gender equity.

Subsequently preoccupation with women's leisure disadvantages was joined by a broader interest in the implications of gender differences. The assumption that existing male patterns were necessarily the best even for men themselves, let alone women, was queried. 'New men's studies' made a problem out of male masculinity and the study of leisure and gender lost some of its former feminine associations. Leisure researchers began to explore not just how the different social roles of men and women led to differences in their leisure opportunities and behaviour, but also how their leisure itself helped to produce and reproduce specific kinds of masculinity and femininity. Which sex was leisure privileged was sometimes regarded as less important than the role of leisure in perpetuating gender divisions together with the advantages and disadvantages for men and women in other spheres of life. A parallel concern was with the ways in which leisure could be used by women and, in practice less frequently, by men to challenge and maybe shatter gender stereotypes.

In the third and, up to now, latest stream of debates on gender and leisure, interest has shifted away from the overall differences between men and women towards variations within each of these groups. Researchers have begun to explore how leisure can be used in the construction of alternative masculinities and femininities. Some writers argue that we are now entering a postmodern era in which, even if the differences between older social categories defined by age, sex and social class are not diminishing, there is a trend towards greater variety within all of these groups. Some believe that this variety is reaching a point where 'male' and 'female' are ceasing to be sociologically useful categories. It is argued that there are now traditional but also several kinds of 'new men', traditional housewives and also various types of modern and postmodern women, and that both genders today have gay and bisexual as well as heterosexual options. Some self-styled post-feminists have claimed that their own, if not all their sisters', liberation from the traditional female role has been largely accomplished, and that the pressing issues now are the types of women that they will become. These claims, needless to say, are contested fiercely.

A danger in such a fast moving field is that knowledge may be jettisoned too quickly. One set of questions is answered and researchers move on to other exciting issues. They do not need to remind each other constantly of what everyone knows. A merit in distinguishing streams in the development of this area is that doing so preserves a hearing for the longer-standing issues. The literature has certainly developed much more rapidly than most men's or women's real lives have changed during the last 15 years or so. The rapid conceptual progress has been possible and necessary not so much to keep pace with a rapidly changing society as to compensate for the decades of neglect. Even if there is a trend towards greater variety in the lives of men and women it does not necessarily follow that the overall differences between them must have diminished. Similarly, the fact that there is much more to say does not deny that women endure leisure disadvantages. There may be no need for research to rediscover these disadvantages repeatedly but there is no reason to believe that they have all disappeared during the last 15 years. The issue may have been superseded in leisure research but this does not make the knowledge either invalid or no longer relevant.

Disadvantaged Women

This is the earliest, continuing, and therefore the most persistent set of arguments in the post-1970s literature on gender and leisure. It is a particularly widespread theme. Women have been portrayed as the disadvantaged sex in some way or another in every country where data on leisure and gender have been gathered (Samuel, 1996). The view that males are leisure privileged is rarely challenged directly though in truth the evidence is less than overwhelming (see below). It has proved far easier to identify the processes that lead one to expect women's leisure to be disadvantaged than to demonstrate that their leisure really is deprived.

Time budget data show that men have more leisure time than women. This is true overall, and when like is compared with like, when men in full-time jobs are compared with their female equivalents and so on. The main reason why women have less leisure time is that they do more housework. Israel appears to be the only country where women overall have as much leisure time as men. This has been explained in terms of the long hours worked by Israeli men in important jobs in which women are very thinly represented (Netz, 1996). Elsewhere men have more leisure time but, it has been argued (see below), especially in respect of women, the quantity of free time that people have at their disposal is less important than the experiences that they derive in fixing the quality of their leisure.

As regards participation in leisure activities, women are under-represented among sport spectators and players, and they drink less alcohol than men. Women's presence in licensed premises has risen, but the male customers are still drinking much more. However, women tend to have the higher rates of participation in home-based leisure, and with the exception of sport and visits to pubs there are few differences in the sexes' involvement in out-of-home leisure activities. They are more or less equally likely to take holidays away from home, day trips to the coast and countryside, and to visit theatres, cinemas, galleries, museums and so on (see Table 4.1). As already acknowledged, these may not be identical experiences for males and females. Nevertheless, it has not been demonstrated that women themselves want to play and watch as much sport as men, or to drink as much beer, or that the quality of their leisure would benefit if they were to do so. Even so, there are impressive and persuasive reasons for believing that women at leisure must be disadvantaged.

Income

First, men earn and have more money to spend on their own leisure. Within both sex groups levels of leisure activity rise when people have jobs, and when their jobs are well paid as opposed to low paid (see Deem, 1986). Even though women's labour market participation has risen steadily since the Second World War, there are still far more full-time housewives than househusbands, and women workers are far more likely than males to be in part-time, low status and low paid jobs. Having money and, just as important, earning money, are important for the spending power that is conferred and for additional reasons. Going out to work widens people's social networks beyond their kin and immediate neighbours. Moreover,

Table 4.1. Gender differences in leisure participation.

		Males %	Females %
Participate in active leisure:			
Age	18–24	65	47
	25–31	58	43
	46–52	28	28
	60–66	20	18
Weekly alcohol consumption:		Units	
Age	18–24	19.3	4.5
	25–31	17.2	4.3
	46–52	13.3	3.7
	60–66	10.3	2.4
18–38 Year olds: previous 2 weeks:		%	
Parks etc.		33	35
Coast		27	27
Cinema, theatre		13	13

Source: Cox *et al.* (1993).

individuals who are earning are likely to feel, and to be regarded by others, as having earned the right to spend on their own pleasures.

It may be argued that personal earnings become irrelevant, or at any rate far less relevant, when men and women live in the same households within which financial resources are pooled, and it is the case that, in general, women's standards of living benefit when they gain a share in a male income (Taylor-Goodby, 1985, 1986). However, the allocation of resources within households is more likely to leave men than women with money on which there is no claim apart from their own pleasures (Dale, 1986; McGlone and Pudney, 1986; Pahl, 1990). Moreover, as explained below, men are more likely to feel that they have the right to indulge their leisure interests, particularly when they have earned the money to do so. Women are more likely to attract censure for squandering family resources when attending bingo and visiting pubs than males who watch and play sport.

Leisure equality is most likely in conditions of economic equality. When husbands and wives both work full-time, and have similar earnings, housework is distributed more equally than in other households and the sexes are likely to retain similar amounts of money to spend on themselves. Male and female higher education students have similar sources and levels of income, similar levels and patterns of leisure participation, and are most likely to engage in joint leisure on equal terms, sharing expenses for example (Furlong *et al.*, 1990). When higher education graduates (and younger school-leavers) commence employment they earn similar amounts, but males tend to receive more training and have the better career prospects (Roberts and Parsell, 1991). With the arrival of children it is still nearly always the mother who interrupts her career. By this time it is usually the male's earnings that are decisive in the size of mortgage that a couple can obtain, and therefore the type of house and district where they can live, and their standard of living more generally. This is the context in which the male's career tends to be given precedence in deciding whether a family will stay put or move to another part of the country for example, and likewise in prioritizing the male's leisure interests.

The division of domestic labour

Women do more housework, a lot more than men. Times are said to be changing in this respect. In the 1970s there was much discussion about dual career families and the emergence of a symmetrical family type in which males and females participated more or less equally in both paid and domestic work (Young and Willmott, 1973). It was argued that in such families males and females would engage in leisure on equal terms (Bell and Healey, 1973). There are certainly more women in employment today than in the past but on average they work for fewer hours, for less money per hour, less continuously and less progressively, and in lower status occupations than men. The sexes are still a long way from symmetry in employment. Time budget studies show that there has been a trend towards housework being shared more equally (Gershuny, 1992). However, Warde and Hetherington (1993) have concluded from their study of 325 Greater Manchester households that the remaining gender differences are so wide that any equalization must have been minor. When wives take paid jobs husbands usually increase their participation in housework but it is the women's total workloads that expand most substantially. And the burden of ensuring that family members with different work schedules are all fed and have laundered clothing ready for their work and play tends to fall on the housewife (Hantrais, 1985).

It has been pointed out that men and women usually do different kinds of household chores. Men take responsibility for particular things, the garden, the car, decorating, and perhaps the washing up, for example. Women's housework, in contrast, is typically unbounded. Women may never be unequivocally off-duty especially when they have young children to care for. Hence the 'What leisure?' question that women often address to researchers and which features in the titles of some of the latter's books. Women's leisure tends to be fragmented and contaminated as they try to watch television or read while attending to the cleaning, the ironing and their children.

Men are the more likely to have, or to claim, the time to do their own things. Barrell and his colleagues (1989) interviewed 24 distance runners and their partners to discover how the participants found the time for their sport. This research encountered many examples of male power as time was bought, taken or somehow made available. Males tend to assume that they have a right to time in which to indulge their own interests, certainly masculine interests such as sport. They assume that their children will be cared for, their meals prepared, and sometimes that their kit will be washed and ironed. Shona Thompson (1990) has observed that women have been incorporated into sport largely by servicing male players, and Margaret Talbot (1979) has argued that servicing the play of other family members, and deriving their own pleasures vicariously by watching and otherwise helping their children and male partners to play, are general features of females' leisure. Girls have often been prepared for such futures during childhood, when they are more likely than their brothers to be expected to assist with housework, in school sport when girls have sat on the sidelines and watched the boys play (Leaman, 1984), and in their own time so much of which has been devoted to cultivating feminine identities. This has commonly taken a great deal of teenage girls' time and money. Once a girl's feminine attractions have caught a male this relationship has often become her main leisure activity and any incompatible or inconvenient personal leisure interests and friendships have been dropped (Sharpe, 1977; Griffin, 1985).

As a result of their leisure socialization when young, females may internalize the servicing role. They may feel that they ought to put their children's and husbands' interests ahead of their own, and feel guilty and selfish when spending time and money on themselves (see Samuel, 1996). Some women seem unable even to express, let alone act on the basis of, their personal desires. So with the arrival of children it is usually the woman's career and leisure interests that are sacrificed. Large-scale survey evidence shows that with the birth of children both mothers and fathers reduce their personal leisure activities, but women change much more than men (Smith, 1987). This is illustrated vividly in Betsy Wearing's (1993) study of 60 Sydney mothers with young babies. They all reported some loss of personal leisure whereas their husbands were said to be preserving their own leisure time, spending and activities.

New technology seems to have been absorbed within existing patriarchal relationships at work and in the home. Vacuum cleaners, washing machines and so on have permitted higher standards and expectations of home care instead of decimating women's domestic work time. Second cars (the first nearly always being claimed by the male) may expand women's mobility but may also be accompanied by the expectation that they will ferry children to school and for leisure.

Socio-cultural regulation

There are social norms, widely shared views and feelings, about the kinds of leisure activities that are appropriate and unsuitable for both men and women, but women are the sex whose lives tend to be the most restricted by these processes. In their study of over 700 Sheffield women, Eileen Green and her colleagues (1990) were repeatedly told that it would not be right or, at any rate, it would not be considered proper and would damage a woman's reputation, if she visited a public house unescorted, or, if she was married, if she went for an evening's dancing with a group of female friends. Noel Parry and Daphne Johnson's 1970s study of 352 18–32 year olds in Hatfield emphasized just how limited women's leisure options could be, especially for married working class women. For some, television was virtually their only leisure activity. Their opportunities to go out were extremely limited. When they went out this was usually as a family or to visit other family members (Parry and Johnson, 1974). It has been considered acceptable for women in groups or singly to go to church, to evening classes or to bingo, but to very few other leisure destinations. In pubs, sports centres and cinemas unescorted females have been assumed to be available for the attention of predatory men. Any women who have looked attractive have been deemed to be 'asking for it'.

This is another respect in which times are supposed to be changing. However, Sue Lees (1986) has described how girls in comprehensive schools can be subjected to daily verbal and even physical abuse by male pupils. Male sexuality continues to control unprotected, that is, unescorted women, by harassment and abuse in most public situations (see Coveney *et al.*, 1984). Women can now seek legal redress for sexual harassment at work but not usually when they are at leisure. Yorganci's (1993) analysis of 149 returned questionnaires from female sports players found that 54% had experienced or knew someone who had been harassed by a (usually male) coach, and 57% of these incidents were considered sexual.

No one claims that all males are personally guilty of sexual harassment. Rather, the claim is that masculine norms condone the conduct and protect the perpetrators from censure. It does not need all or even most men to be harassers to make most women feel vulnerable. Fear of being out of home alone, especially at night, was mentioned time and again by the women in the Sheffield study (Green *et al.*, 1990). Males can visit cinemas and public houses, and play sport without fearing similar harassment (though they are actually more likely than women to become victims of male violence), and it is not only young women who feel vulnerable. Mason's (1988) study of 50–70 year old women found that their leisure was restricted by lack of free time, money and fear of harassment.

It may not be so much that males' attitudes and behaviour have changed in recent years as women's former acquiescence. The older women in Alan Tomlinson's (1979) study explained that they had willingly given up the leisure of childhood and youth in exchange for different ways of life which were filled by family and community work. Dorothy Hobson's (1979) study of young working class married women, conducted at roughly the same time as Alan Tomlinson's, found that these younger subjects had misgivings about the passing of the period between leaving school and settling down when they had been able to see their own friends regularly and enjoy a good night out.

A further aspect of the cultural regulation of women's leisure is that their interests and activities tend to be devalued. Anything done mainly by women, playing bingo for example, is likely to be dismissed as frivolous. This is a further manifestation of male cultural power. O'Connor and Boyle (1993) have argued that television soap operas and sport offer very similar gratifications to their audiences: gender role models, emotional expression and material for small talk. Yet sport is treated as the more worthy, serious business. The tastes of its followers are legitimized by the daily treatment of sport in the media. The sums of money that men spend in following their teams are not condemned. The cash that women are said to fritter away on bingo, and their alleged naivety in becoming hooked on soaps, probably say more about male cultural power than the respective motivations and gratifications of sport and bingo players and the followers of television light entertainment.

Leisure provisions

Most leisure providers are male. They occupy most of the power positions in the public authorities and private sector businesses that cater for leisure. In so far as these are run mainly by men it is possibly only to be expected that the providers will cater primarily for men. As yet there is no hard evidence that women in power positions act differently in politics, leisure or other businesses. Maybe the exceptional women who have 'got on' needed to become 'one of the boys'. However, it is difficult to believe that the magazine and video shops would continue to display pornography, that women's sports and women-only sessions in sports centres would not find it easier to claim prime time, that sportswomen would remain as vulnerable to harassment, or that the current bar room cultures would continue to be tolerated if most of the power positions in all the relevant organizations were occupied by females.

Sheila Scraton (1987, 1992) has argued that girls are eased out of sport not so much because they are unable to enjoy the activities themselves than by the surrounding masculine culture that conflicts with girls' preferred feminine identities. Would this culture remain so pervasive if most of the power in sport was wielded by women? Would the portrayal of women in the media remain unchanged if most television stations, newspapers and magazines were run by women? It is possible only to ask these questions rhetorically because it is impossible to experiment. Universities, colleges and schools have maintained acres of playing fields and other expensive sports facilities which have been used mainly by male pupils and students. Would this have happened with so little controversy if women had been spending the money? Or would there have been steps towards equivalent spending on the interests of female pupils and students?

Once again, change is said to be underway with more and more women breaking through former glass ceilings, and with public and private sector leisure providers becoming more sensitive to women's tastes, aspirations and spending power. However, in the public sector there have been signs of such changes grinding to a halt. Feminist demands may have been most influential in the 1970s when welfare and community development ideologies were fashionable (Yule, 1997a, 1997b). In the 1990s market-oriented thinking has been ascendent. Local authority leisure services departments have been reducing the number of women's officers, and special provisions for women in sports centres have been cut back (Aitchison, 1997).

Female writers have warned of the dangers of requesting better leisure provisions for women while these demands have to be addressed to powerful men (see Green *et al.*, 1990). They fear that any targeted provisions will be on the basis of men's conceptions of women's needs and entitlements: mother and toddler sessions, and women's sports at off-peak times for example. Women players often value sport because it provides them with time and space away from their homes, husbands and children (Roberts and Brodie, 1992). It has been argued that the preconditions for equitable leisure provision are that women organize to express and press their own demands, and penetrate the power positions in leisure provision organizations thereby becoming the decision takers. These preconditions, needless to say, could arise only if the other causes of women's leisure disadvantages were removed: their second class status in employment, their domestic workloads, and the less tangible social and cultural obstacles to equal leisure opportunities.

Gender Differences

Preoccupation with women's disadvantages may be justified by their severity but can still produce a rather distorted view of women's leisure. After all, some gender differences may be just differences rather than inequalities. In 1982 Sarah Gregory urged greater attention to women's leisure achievements instead of repeatedly deploring their failure to escape from male oppression. Later writers, having noted that the conventional male concept of leisure fails to match the realities of most women's lives, have embarked upon alternative conceptualizations. For example, Lenskyj (1988) has argued that there are three major contrasts between men's and women's leisure orientations. First, she argues that women tend to be task rather than time oriented; that women's ways of life lead them to associate leisure with enjoying particular tasks rather than discrete periods of time. Second, she argues that women tend to be altruistic rather than self-centred and derive genuine satisfaction by contributing to the well-being of others. Third, she argues that women are less concerned with the quantity than the quality of their leisure activities and experiences. The sexes' different approaches to leisure appear to be rooted in broader feminine and masculine value orientations. In Western cultures women tend to be the more compassionate sex (more sensitive to the well-being of others), more concerned to identify inner meaning and purpose in their own lives, and generally less materialist and competitive than men (see Beutel and Marini, 1995). Lenskyj's contrasts between men's and women's leisure make the former's priorities – grabbing as much time as possible for themselves, and filling this time with as many activities as possible – appear distinctly unattractive. Table 4.2 lists the characteristics of women's leisure identified by Lenskyj, and other writers whose work is reviewed below, together with the masculine opposites. It is doubtful whether anyone would conclude from the characteristics listed that males are clearly leisure advantaged.

Other writers have echoed many of Lenskyj's points about the characteristics of women's leisure. For example, Betsy and Stephen Wearing (1988) suggest that women are less likely than men to associate leisure with

Table 4.2. Stereotypes of feminine and masculine leisure.

Feminine	Masculine
A kind of experience	A type of time and/or activity
Embedded in other activities	Separate part of life
Task oriented	Time oriented
Relationship based	Activity based
Seek quality experience	Seek maximum activity
Cooperative	Competitive
Self-generate	Purchase

given times and activities, and more likely to describe any times and activities as leisure if they are freely chosen and yield self-actualization (self-development and expression). Approaching women's leisure in these terms is said to allow their portrayal as enterprising and enabling women rather than passive victims. Karla Henderson and her colleagues (1989) have also noted the importance of choice and enjoyment (the type of experience) in women's ideas about leisure, an implication being that leisure may be experienced within, and need not be separate from other areas of women's lives including their paid and unpaid work. Hence the case for a feminist concept of leisure or, at any rate, one that is equally applicable to men's and women's experiences. The 24 women in rural Australia who were interviewed by Penny Davidson (1996) had varied views on holidays. Some viewed them as work whereas others regarded going away as an opportunity to spend 'quality time' with their families. Davidson uses this evidence to query the validity of defining leisure as a part of life. She argues that leisure is '. . . an experience that is defined as much by relationship and contribution to self-identity as it is to reduced pressure and pleasure' (p. 102).

It has been argued that the study of women's leisure can benefit from specifically feminist methods such as those employed by Davidson (see also Green *et al.*, 1990). Many feminist writers have argued that qualitative methods are vastly superior to quantitative techniques in exploring women's lives. The methods developed by Luise Eichenbaum and Susie Orbach (1984) in exploring what women really want were not devised with leisure specifically in mind, but they could be applied in this area. In their work Eichenbaum and Orbach used individual and group therapeutic sessions to prompt their women subjects to recognize and express their authentic desires and feelings. Needless to say, it can be disputed that these, or any other research methods, are specifically feminist.

Strengths of women's leisure

Focusing on gender differences and approaching women's leisure from a feminine or feminist perspective has several advantages. First, as we have seen, it allows the distinctive strengths, as well as any disadvantages inherent in their types of leisure, to be identified. Betsy and Stephen Wearing's (1988) claim that focusing on women's choices and experiences can highlight their initiative and self-direction has already been mentioned. Other writers have praised women's ability to make full use of time (often doing two or more things simultaneously), their resourcefulness in deriving leisure experiences from unpromising situations, their ability to adapt to change, and their inter-personal skills which allow women to confide their own feelings and facilitate others in expressing their's (Gregory, 1982). These alleged strengths of women's leisure encompass the features that are said to make women the skilled time managers whom men will need to emulate to adjust to ongoing changes in the world of work (see Chapter 3). Instead of trying to overcome the barriers to male leisure, it has been argued that truly woman-friendly leisure policies and provisions will take their lead from women's existing wants (Gregory, 1982).

However, the fact that families can be a source of satisfying and productive leisure (Tomlinson, 1979) does not necessarily mean that women should remain reconciled to deriving leisure experiences in this way. Rosemary Deem (1982) has argued forcefully that most of the alleged strengths of women's leisure arise from making the best of heavily constrained and basically disadvantaged situations. She argues that the main beneficiaries of most of the differences between men's and women's leisure are in fact men. Deem does not want to force anyone to be masculine but insists that women should have the same range of leisure options as males. This, she argues, will only become possible when women have jobs, economic independence, self-confidence, private transport, when they themselves and others recognize that they have a legitimate right to pursue personal leisure interests, and when women have their own support networks, friends who may be either male or female, who are able to assist individual women in doing just this (Deem, 1986).

Leisure and gender role socialization

A second merit in focusing on gender differences is in exploring the myriad ways in which leisure may not merely reflect but also affect the types of people that individuals become. Girls are given dolls and prams and are instructed to be neat and tidy. Boys are set to play with footballs and soldiers and are excused if they are rough and dirty. Inappropriate play by either sex is discouraged by most parents. Teenage boys are given the freedom to roam neighbourhood streets and city centres where they learn to claim public space. Girls are more likely to be required to do housework and are generally more protected. Parents often insist on escorting their daughters to and from leisure destinations, and want to know where they are going and precisely when they will be back.

It is largely through play that people first learn to act, and to feel, masculine and feminine. As previously mentioned, a great deal of teenage girls' leisure is devoted to the construction of femininity. Likewise teenage boys spend a great deal of time playing at being men. Among girls, choosing the appropriate clothes and hairstyles, preparing for a night out, then discussing what happened afterwards, can take a great deal of time, effort and money. When teenage girls have become involved with males they have often been expected to subordinate any personal interests and former friends to the new relationship, thereby becoming accustomed to the servicing role (Sharpe, 1977; Griffin, 1985).

The consequences of these leisure practices are likely to extend well beyond leisure itself. Eileen Green and her colleagues (1990) have stressed the importance of leisure as a site where patriarchal relationships are reproduced and became hegemonic, meaning that they are made to appear 'just natural'. If people behave in masculine and feminine ways during their leisure, when they have the maximum scope for choice, the apparent implication is that they are expressing their basic natures rather than playing roles imposed on them by the wider society.

Young people often resist the roles that adults cast for them. Teenage boys and girls often rebel against parental and teachers' authority. What is interesting for present purposes is the frequency with which masculinity and femininity are used by boys and girls respectively to assert their autonomy. Paul Willis's *Learning to Labour* (1977) describes graphically how lower stream working class boys may reject the 'sissy' aims that their teachers urge upon them and insist that they intend to become real men who do real work. Girls often transform school into social life by using toilets and other spaces to create their own time to talk, use make-up and so on (Griffiths, 1995). Young people thereby commit themselves, ostensibly entirely voluntarily, to conventional male and female roles in their future family and working lives.

Challenging stereotypes

Another virtue of the gender differences perspective is that while drawing attention to how leisure can sometimes socialize people into, it can be equally sensitive to the ways in which leisure may be used to challenge and resist, conventional gender roles and divisions. A third possibility is that people may use conventional gender roles for their own purposes. Jennifer Hargreaves (1994) points out that some female athletes do just this when surrounded by sexualized imagery. This may benefit event sponsors and promoters but the female athletes are not necessarily being exploited reluctantly. Some use their sexuality deliberately and actively to promote their own careers. The same is no doubt true of many actresses and actors, as well as men and women in relatively mundane occupations. The point is that neither sex need passively conform; individuals usually either embrace actively or challenge (in some way and to some degree) the stereotypes into which their society appears intent on slotting them.

Vivienne Griffiths (1995) has argued that teenage girls' leisure is not devoted only to making themselves attractive to boys, and that they do not always sacrifice their own interests and existing friends for the sake of a mere male. Her study of 12–16 year olds confirmed the restrictions to which girls in this age group are usually subject by their parents and the general threat of male harassment. Yet Griffiths found that the girls she studied were succeeding in constructing busy leisure lives. They were finding time and space for themselves in clubs, bedrooms and elsewhere in their homes. Griffiths also emphasizes that the popularity of disco dancing among teenage girls is not solely on account of the opportunity to display themselves to male audiences. Many teenage girls prefer to stay and dance with girl friends. Dancing ability can be a source of self-esteem and peer group admiration, and straightforward sensual pleasure, with or without male admirers (see also Brennan, 1993).

It is probably the exception rather than the rule when leisure is used to challenge gender roles and divisions. However, there are plenty of examples in recent times of young and older women using leisure to contest patriarchy. Margaret Talbot (1990) and Sheila Scraton (1992) have both drawn attention to how sportswomen can challenge orthodox views of femininity. Women can use sport to prove to themselves and others that they can be strong, independent and competitive. Eileen Green (1998) claims that leisure contexts are important for women, especially those contexts that create space in which women can talk to one another. She argues that through leisure talk women can develop and consolidate preferred identities with which to challenge conventional stereotypes.

Perhaps this is how women have developed the confidence to claim formerly male public space in sports centres, pubs and city precincts. Leisure can be used to defy gender stereotypes without the risk of personal costs comparable with losing a job or being unable to obtain employment. But having discovered that gender roles can crumble, the moment that sufficient numbers and sometimes just one solitary individual decide to defy them, the lesson may well be transported into other domains.

Vulnerable males

Interest in all gender differences rather than just women's lives and restrictions forces recognition that masculinity is imposed upon males just as females are socialized into femininity. Many males have always submitted to conventional masculinity with reluctance. Some have not enjoyed tough and competitive sports, having to play the hunter-aggressor in heterosexual encounters, or suppressing any inclination to display their emotions in public. The desire of some males to explore alternatives has helped to usher in the 'new men's studies' (see Morgan, 1992; Nardi, 1992). Both the field of research, and many of the men who are involved, can be described as 'new'.

Several recent studies have drawn attention to the frustrations and vulnerability created for some males by the social expectation that they should be conventionally masculine. Wright's (1994) interviews and observations among 58 14–16 year old working class males in Glasgow found that, although their group talk objectified females and treated them primarily as vehicles for male gratification, privately many of the teenage boys expressed frustration with the passive role to which girls were expected to conform, and with which, in the boys' experience, they usually did conform.

Valerie Hey's (1986) analysis of the two sexes' experiences and behaviour in public houses underlines the harassment to which female customers and staff have been routinely subjected. However, she argues that much of this stems from male vulnerability. Hey notes the importance that men attach to their reputations among other male regulars. Some men feel that they need to go out with their mates regularly and drink fulsomely in order to prove that they are not 'under the thumb'. Relationships within these male groups are not emotionally intimate. The men talk about sport and cars but not about their private lives except in jocular or otherwise superficial terms. This is the world of male small talk that women can find impossible to penetrate in leisure or work situations. But because so little of their private, emotional lives is known to their male friends, individuals' public identities are liable to be shattered in any face-to-face encounters with women. Hence, Hey argues, the men's need to police females into well-defined subordinate roles.

Maybe the injured male should not be writ too large. Rosemary Deem (1982, 1986) insists that men have been the net beneficiaries of gender differences in employment, the family and leisure. There has been much more servicing of men's leisure by women than vice versa. Men have been able to walk the streets and parks without fear of predatory females. Even so, both sexes could gain from change. Duncombe and Marsden's (1993) highly original study of couples draws attention to how men and women tend to talk and think emotionally in quite different ways. Women are taught to express their feelings whereas men have been expected to maintain stiff upper lips. So women complain about their partners' limited emotional participation while their men, who do not want to burden anyone else with their feelings, either cannot understand or, even if they do, are incapable of response. In some ways, Western forms of masculinity and femininity appear to make men and women emotionally incompatible (Nare, 1996).

Up to now it has been women who have been more likely to question and challenge gender differences and divisions, but men are becoming increasingly sympathetic. Changes in the occupational structures of advanced industrial societies are destroying swathes of masculine jobs. Self-confident and assertive females have been delaying marriage, then leaving their partners in increasing numbers. Females have begun to out-perform males at all levels in education, and in the competition for jobs among Britain's university graduates. Women's tastes have been crucial in the conversion of pubs into wine bars. Young women are now even matching young men's presence in sports centres (Department for Education, 1995). A Dutch panel study of young people and their parents found that young males and females alike wanted different lives and relationships from those of their parents (du Bois-Reymond *et al.*, 1995). The young people of both sexes in this study wanted to combine labour market careers with participant parenthood. And in searching for new roles the Dutch young people had the support and encouragement of their parents, especially their mothers.

Masculinities and Femininities

Instead of making broad contrasts between men's and women's leisure, the 1990s has seen a trend towards investigating and stressing the variety within each sex. The significance of challenges to conventional gender stereotypes has been reassessed. Have assertive women been changing what femininity means for everyone or creating an alternative just for some women?

Post-feminism

Self-styled post-feminists have argued that women in the 1990s have been able to benefit from the wider opportunities won through women's struggles in the 1970s and 1980s. They draw attention to women's strides in education and the labour market, and, by some women, in domestic life and leisure. Everyone realizes that not all women's lives have been transformed, but when women are not benefiting in any obvious way from their sex's new opportunities this may be because they do not wish to do so (see Hakim, 1995, but see also Ginn *et al.*, 1996). Traditional femininity has remained an option to which some women may not want an alternative for themselves. Feminism, the label at any rate, has ebbed in popularity among women. Young women seem to be increasingly individualistic, less interested in bonding in a movement and campaigning against the restrictions that remain than in using the opportunities that have become available for their sex.

Choice

Greater variety is said to be an outcome of this situation. There are still many full-time housewives but there are increasing numbers of women pursuing full-time continuous careers in the labour market. More women than in previous generations share the breadwinner role, but there are also more single parents. Is it possible any longer to generalize about the situation of women? This greater variety extends into leisure. More women are playing sport and enjoying nights out in pubs and clubs in all-female groups, but others have not been part of these trends. There is also greater variety among men. More are living singly and more are in dual career partnerships than formerly, but the number of sole breadwinners has not shrunk to insignificance. More men than in the past, albeit still a small minority, are taking career breaks and opting for reduced hours of work so that they can play a larger role in child care and, they hope, lead more balanced, better quality lives (Walton, 1996). The variety is not entirely new, but some long-standing variations are now receiving closer attention because they are posing new problems. For example, Mac an Ghail (1996) has emphasized how young middle and working class males have always expressed masculinity in rather different ways, but then goes on to explain how working class masculinity is currently in crisis due to the disappearance of so many of the manual jobs that the males formerly entered. In this situation the males may either devise alternative ways of expressing and confirming traditional working class masculinity, or redefine what it means to be masculine. Against this backcloth it has been argued that instead of emphasizing the differences between all males on the one hand and all females on the other, the focus should be on how various groups of males and females use their leisure (and education and labour market opportunities as well) to construct or select from the variety of masculine and feminine identities that are now available. Sexuality is just one example of how, it is sometimes suggested, males and females now have options; to be gay, lesbian or bisexual rather than compulsorily heterosexual.

Traditional, modern and postmodern

Lia Karsten's (1995) study of 40 20–45 year old Dutch women is a good example of this trend. She used diaries, interviews and questionnaires to distinguish three types of women in her sample. First there were 'traditional' females (a declining group). They did not have paid jobs and therefore had ample free time but did not claim this as their own. Rather, they entwined virtually all their leisure with caring and servicing. Two growing types of women were also identified. One was 'modern' women who were usually well-educated and had paid jobs, but these were more likely to be part-time than full-time. However, what identified these women as a leisure group was that despite earning their own money they subordinated their leisure to the needs of their families. Karsten's third group was composed of 'individualized' women. Some were single and had full-time jobs. Others were married and if they had jobs these were usually part-time. Full-time and part-time employment, and living singly and being married, were all proving compatible with individualized leisure when women were determined to demarcate their own free time. Karsten suggests that which of the new types becomes dominant in the future will depend on the wishes of women and also on the attitudes of men and the outcomes of 'negotiations' between the sexes. Her key point, for present purposes, is that sex roles have ceased to be rigid fixtures to which all men and women are required or even expected to conform by 'society in general'.

Another Dutch study (see te Kloetze, 1998) has identified similar changes to those discussed by Karsten. This second study was conducted over a 20 year period in two Dutch villages, and noted a movement from traditional, to modern, to postmodern family patterns. Traditional families were embedded in neighbourhood and kinship relationships which presented and enforced different ways of life for the sexes. Modern families escaped from these communal constraints, and emphasized conjugal togetherness and sharing (of paid work, child care and leisure). In the latest type of postmodern family the partners had greater respect for each other's individual rights, including the right to pursue independent leisure interests. However, there were limits to how far even postmodern couples were prepared to pursue individualization. For example, they were nearly all opposed to separate holidays.

This stress on the sexes' new found ability to choose and construct various types of masculinity and femininity has encountered some hostile treatment. It has been pointed out that variety among men and women is not new. Nor did it escape earlier writers' attention (see for example Talbot, 1979). However, earlier writers linked the leisure differences among men and women to their ages, ethnic groups and, in particular, their social classes, rather than either males' or females' lifestyle choices. And the earlier writers invariably stressed the differences between, and the advantages and disadvantages shared by, all men and all women respectively.

The limits to change

Some argue that these points should continue to be stressed because there are still some glaring overall differences between the leisure of women and men. Critics of post-feminism argue that only a minority of females have been able to benefit from the new opportunities to evade former constraints. Sheila Scraton (1994) insists that women as a group are still heavily disadvantaged in the labour market, in the domestic division of labour, and by the threat of male violence. Women's opportunities, she argues, whether in the labour market or at leisure, are still governed primarily by their life chances rather than their own lifestyle choices. Coppock and her co-authors (1995) have queried whether women's positions have changed fundamentally in any way in recent decades. They argue that the liberal reform agenda has patently failed, that male power is still firmly in place, and construe post-feminism as a male inspired backlash against feminism's very limited achievements.

Tess Kay (1996) has argued that there have been changes for some women, mainly a highly educated minority who have been entering management and professional jobs. These women's male partners, if they have any, are most likely to have similar occupations. There has been a trend towards such women working full-time and continuously and, if they have children, paying for child care. Such women's lifestyles are increasingly resembling their male partners'. However, Kay shows that this trend has been confined to the upper socio-economic groups. Women with children aged under five in Registrar General's Social Class I are 30 times more likely to work full-time than women in Social Class V (unskilled manual). Kay argues that the lifestyles of the higher socio-economic groups cannot spread downwards because there are only a limited number of top jobs, and gender segmentation remains strong, generally to women's disadvantage, in lower level employment. The net result is a narrowing of gender differences at the top of the socio-economic hierarchy alongside a widening of social class differences within each sex. Within the working class there has been no coming together of the sexes' opportunities and lifestyles. Indeed, in some parts of the UK traditional gender differences are likely to have strengthened as youth unemployment has prevented males from becoming providers while, among females, it is associated with early (usually single) motherhood (see Roberts *et al.*, 1990).

So the older arguments about women's disadvantages, and the general differences between men's and women's leisure, remain of current rather than just historical interest. These issues have been joined by others, but they have not been superseded completely by social change.

Unfinished Business

It would be premature to declare conceptual progress complete, even for the time being, in the study of leisure and gender. The battery of criticism of 'malestream' research that launched the still flowing waves of enquiries has not become wholly outdated. There is no longer a crude shortage of research into women's leisure but studies of men's leisure that explore the construction and implications of their masculinity, or masculinities, are still occasional and usually speculative and anecdotal. Perhaps male researchers are still conditioned by their conventional sex role and are inhibited by the unmasculinity of reflecting on their own, and questioning other men about, sexual identities. There is a persistent imbalance between research on male subjects which focuses on the implications of their ages, jobs and social class positions, and the studies of women that nearly always dwell on their experiences as women. Large-scale investigations that compare the sexes still use 'male' definitions of leisure and research methods. Concepts and methods of enquiry that are said to be more suited to women's lives are still most likely to be used when women are the only subjects. Despite the spread of interest in experiential concepts of leisure (see Chapter 6), these have not been refined and systematized to allow their incorporation in large-scale comparative studies in which men's and women's lives could be treated in an even-handed way.

Sylvia Walby (1988) has identified four stages in the development of research and writing on women and politics. In the first stage women were overlooked. In the second stage this neglect was criticized. In the third stage the study of women was added to political science as a special area. In the fourth stage, still incomplete when Walby was writing, mainstream theories and concepts are modified and the incorporation of women is thereby completed. Such a final stage has still to be tackled in the study of leisure.

When this stage is reached some of the outstanding issues may prove incapable of settlement to everyone's satisfaction. Even incorrigible antifeminists have been forced by the weight of evidence and argument to concede that the neglect of women and gender in leisure research prior to the 1980s had to be rectified. However, not everyone will be content if (as in this book) gender is simply added to the list of factors including occupation, age, ethnic group and place of residence, which are acknowledged as being associated with particular constraints and opportunities and, therefore, uses of leisure. Up to now this has been the mainstream response but it is a minimalist answer to the calls for a fundamental reconstruction of 'malestream' leisure theory and research.

Patriarchy is not peculiar to capitalist societies, or even to modern societies more generally. It is widely agreed today that patriarchy preceded modernity, but it is hugely controversial, though not wholly implausible, to argue that patriarchy governed modernization which, according to this view, was and remains a thoroughly patriarchal project. The claim is that capitalism and industrialism (and state socialism) were created by and for men who have monopolized positions in public life and asserted their right to perform paid productive work thereby 'earning' their leisure. This has been possible for men, this argument runs, only at the expense of conscripting women into vital reproductive work, usually unpaid, which results in their leisure being different if, indeed, the term has any application to their lives. Rather than being one of many factors related to its use, gender is highlighted as *the* force which has produced and shaped present-day leisure. Differences among women and among men are, of course, acknowledged but it is argued that these differences are nearly always within masculine and feminine forms between which there is a qualitative gulf. If this analysis is accepted the apparent gender neutrality of the definition of leisure with which this book opened would have to be rejected as another product of a malestream which keeps women's experiences invisible or marginal while pretending to be even-handed. Critics will argue that the gendered character of leisure should have been recognized at the outset in which case everything that followed – about the growth of leisure and the implications of different types of employment for example – would need to be reappraised and rewritten.

A good reason for declining the invitation to start again is that, in my view, modernization was not driven *primarily* by masculinity but by economic classes and scientific rationality. Pre-modern societies were gendered and patriarchal, so most modernizers were men, and established gender relations were carried forward and reshaped by the new social order. But the new social order's distinctive forms – industrialism and the roles of employers and workers for example – have always been capable of enactment by either males or females. The type of leisure that is available to employees in societies with market economies and political constitutions that confer freedom of association, can be accessed by males and females, on equal terms and in equal amounts, if that is how the sexes decide to order their relationships. Sex could, in principle, be separated from all the social distinctions relevant to modern leisure opportunities without scrapping the modern organization of work or the market economy. Up to now this has not happened. Maybe men and their masculinity have been a big obstacle, but the barrier has not been a masculinity necessarily inherent in modern institutions. Industrialism and capitalism will tolerate a variety of gender arrangements. This is why gender roles have been able to change during the modern era and why it is possible for gender relations to vary between sections of the population. Within some sections, among young single people for instance, the sexes' uses of leisure are more similar than among other groups.

The distinctive features of leisure in the modern world do not arise from patriarchy. If this was the case the modern world's leisure would not be so distinctive. It is access and aspirations, not the character of leisure itself, which are gendered. Up to now the division of labour by gender has entailed major differences in males' and females' leisure opportunities and aspirations, but it is not the core institutions of modernity that dictate that this must be so. This is why gender has to be recognized as an important source of leisure inequalities and differences but does not qualify as the foundation, or merit treatment as the principal shaping force and master key to understanding modern leisure.

Chapter 5

The Life Course

Rhona and Robert Rapoport's *Leisure and the Family Life Cycle* was published in 1975 and remains the sole British book (apart from edited collections) to couple leisure and the family or ageing in its title. Leisure and the family is covered in numerous books with a broader scope but this is not the same as adopting the family as the organizing frame. There are many books on the leisure of specific age groups, especially young people, but this is not the same as examining the lifelong links between age and leisure. The limited attention that they have paid to the family and ageing surprises even leisure researchers themselves since they are all aware that the home is the place where, and other family members are the people with whom, leisure time is most likely to be spent. This has been revealed ever since leisure research began. It is also well known that age is strongly related to leisure behaviour (see Table 5.1). There is a general decline with age in the range and frequency of participation in out-of-home leisure. In some activities – playing sport and drinking alcohol for example – this decline is very steep. Exceptions to this general rule – visiting parks, churches and bingo – stand out vividly. As people grow older, more and more leisure time is spent

Table 5.1. Use of free time in 1995 by age.

	16–24 (%)	25–34 (%)	34–44 (%)	45–59 (%)	60+ (%)
Mainly at home	42	50	52	58	67
Mainly away	32	25	23	18	14
Socializing	24	22	23	20	15
Doing nothing	2	3	3	5	4
Mean hours per week	40	37	33	40	52

Source: Martin and Mason (1998).

at home. Leisure researchers will agree that the family and age are important variables and also that they are difficult to tackle.

One problem with the family is that it is difficult to disentangle the effects of dwellings, their contents, family composition and family relationships. Another difficulty is that the family is not the same kind of group for everyone. There are different kinds of families, and the same family can be a rather different setting for different members, for men and women, and adults and children, for example.

The basic problem with ageing is that it happens to everyone and the members of different age groups are in fact the same people at different stages in their lives. Sex, in contrast, is a lifelong attribute and social class positions are usually occupied for the greater part of, if not the entire life course. Being of a certain age cannot exert the same kind of lifelong influence. Studying the effects of ageing requires the investigation of changes over time, which takes time, and it proves difficult to separate the effects of historical change, rising standards of living for example, biological ageing, and changes in the actors' social roles. By the time the disentangling has been completed, the detailed findings are likely to have become outdated.

From Life Cycle to Life Course

The life cycle

Rhona and Robert Rapoport's solution to some of these problems was to adopt the life cycle as their central concept. Following extended interviews with a sample of London families they argued that during their lifetimes most individuals were experiencing three main careers: work careers, family careers and leisure careers. The Rapoports illustrated how these careers typically interacted to create successive life stages, each with its own preoccupations, usually arising from the actors' work and family situations, which gave rise to particular interests which, in turn, led to characteristic uses of leisure. They argued that adolescents were typically preoccupied with identity problems. 'Who am I?' was said to be the preoccupation characteristic of this age group, and in order to discover themselves young people were said to seek a variety of leisure experiences. Hence their high participation rates in out-of-home recreation. Young adults, in contrast, were said to be preoccupied with embarking upon work careers and family formation, so their leisure became more focused around particular partners and, subsequently, more home centred. Afterwards individuals were said to become preoccupied with establishing themselves in their careers and families, and their leisure became even more home and family centred, usually in ways that strengthened family bonds. The Rapoports argued that life cycle transitions associated with starting employment, getting married and becoming a parent, for example, were important moments for leisure because it was at these points in life that patterns of day-to-day living were likely to unfreeze and become available for reconstruction.

Needless to say, the Rapoports recognized that each life cycle stage had a rather different significance for males and females, and that there were important social class differences throughout. However, it appeared that most people were joining particular social class and gender groups early in life, then following the life cycles characteristic of their groups. It was as if the members of each group were being herded into public transport vehicles when young, after which each group would travel through successive life stages together sharing typical experiences throughout (see Berger *et al.*, 1993). The members of any age group could look ahead, see what had happened to the cohorts who had been in their own situations some years previously, and thereby see how their own lives were most likely to unfold in employment, the family and leisure. In 1975 the Rapoports' use of the life cycle concept was uncontroversial. Its introduction to leisure research was generally applauded. At that time family relationships usually lasted for life, and there were clear and firm links between age and most other social positions.

Destandardization

Since then life seems to have stopped going round and round in cycles. The social sciences are now littered with references to the life course being destandardized. This is another example of the recent loosening of older structures. Labour markets have become more flexible, occupational careers have become less secure, risks of unemployment have risen and work schedules have become more varied. Alongside these economic trends neighbourhoods have become less close-knit than they once were, families have become less stable, and researchers now recognize a variety of masculinities and femininities. Major life events are no long as closely linked to given ages as in the past. People today often retire from their main occupations when, in other respects, they are at the peak of adulthood. People in mid-life have been restarting careers, sometimes by returning to full-time education or training. More people in their 30s and 40s are experiencing the dissolution of their domestic partnerships and rejoining the singles scenes.

An upshot is that there is greater variety within all age groups. It has become more hazardous to try to generalize about the circumstances and behaviour of people at any given life stage. Among women aged 25–40, for example, there are still some traditional housewives, but more than formerly are in dual career partnerships, more than formerly are also single parents, and more than formerly are also single and childless. Age has become a less useful predictor. By age 25 there are still many young adults who have established themselves in good jobs, but many others have poor jobs, many have obtained most of their labour market experience in schemes and unemployment, and more than in the past are still full-time students. Some writers regard the all-round destandardization and flexibilization, which have been spreading, as features of a new postmodern condition (see Chapter 8).

The life course

There is still a life course. Age has not become socially meaningless. It remains a powerful basis for social discrimination in situations ranging from labour markets to discos. People's behaviour and experiences early in life still have profound implications for their subsequent opportunities. For example, it remains as important as ever to become as well educated and qualified as possible when young in order to maximize one's chances of a high rising career in management or the professions. The change is that educational success when young no longer guarantees entry to such a career, and entry no longer guarantees that the career will last for life. It is still as necessary as ever to have children to stand any chance of becoming a grandparent and receiving economic and social support from one's descendants in later life. However, parents have become less able to rely on their family links remaining strong.

It is worth stressing that the destandardization of the life course is a trend, not an absolute state. The trend explains and justifies the growth of interest in groups such as single parents rather than all parents, and the early retired rather than all older people. Researchers have become more sensitive to variations within all age groups (as is the case with gender). But the full effects of the recent destandardization of the life course will become evident only during the lifetimes of current and future generations of young adults. They have been exposed immediately and directly to the full force of the trends, and it is they who will experience marriage dissolution levels of 40% or more if current rates continue. Up to now no cohort has gone through life and experienced this degree of family flexibility. However, youth has already become a different kind of life stage than in the recent past for the majority of young people. As yet the same cannot be said for later life stages. This is now an additional consideration to those which, in the past, inspired special interest in young people's lives. There are some obvious continuities. Young people are still the age group that is most receptive to the latest sounds and fashions. They still have the highest participation rates across out-of-home leisure in general. But are these continuities basic or just superficial?

Youth Cultures

Youth's new condition

'New' means post-1970s. There are several stark contrasts with young people's earlier situations. First, the life stage has been prolonged. Perhaps it is more accurate to say that its length has become more varied. There are still some young people entering employment at age 16. Some teenagers still marry and become parents. But the typical ages at which young adults cross all these thresholds have risen. This is partly because jobs have become more difficult to obtain amidst the higher levels of unemployment that have followed the 30 glorious post-war years. Simultaneously, young people have become keener to remain in education so as to become as well qualified as possible. A desire for qualifications was once considered middle class whereas nowadays parents and young people from all social classes recognize the advantages, especially in tight labour markets, of entering as highly qualified as possible. Family transitions have slowed down partly because transitions into employment have been delayed. However, young women's control of their own fertility, their improved labour market opportunities, some parents' greater ability to support their children through prolonged transitions, and young people's own worries that their partnerships may dissolve have all contributed to this slow-down (Hobcraft and Kiernan, 1995; Irwin, 1995).

There has been no upward movement in the characteristic ages at which youth commences. Sixteen has remained the statutory school-leaving age in Britain since 1972. By age 16 the majority of young people have adopted some adult leisure practices. For example, the majority are drinking alcohol regularly. The age of first full sexual intercourse is probably as good as any other single indicator of how slowly or quickly young people are establishing independent lifestyles, and this age, at present typically 16, has not risen. Nor has there been an all-round upward movement in the ages at which young people first leave their parents' homes to live independently. The mean age has actually fallen in Britain as more young people have progressed into higher education (Jones, 1995). Most young people now experience an intermediate stage between leaving their parents' homes and marrying. Until the 1960s, for most young people, these transitions occurred simultaneously (see Leonard, 1980) whereas today most marrying couples are already living at the same address. Cohabitation has been normalized. Travelling is another way in which some young adults fill the life space created by the prolongation of their transitions.

Second, young people's biographies have been individualized. This trend has already been mentioned in earlier chapters. Among young people it has spread partly due to the variety of courses in post-compulsory education, training schemes, part-time and temporary jobs, and periods of unemployment, which they now experience. It is also due to the contraction of the major firms and industries that once dominated many local labour markets. There was never a village where absolutely every boy went down the pit or every girl into the cotton mill, but in many parts of the country there used to be main types of employment into which most males or females, with specific educational backgrounds, would progress. Individualization is also a product of the break-up of neighbourhoods that were once knit together by their residents' lifelong acquaintance. These trends have made it more difficult to conduct youth ethnography in the traditional ways. It was once possible for fieldworkers to make contact with (usually male) peer groups in given localities and emerge with portraits of their typical backgrounds, attitudes, lifestyles and futures. These groups are no longer present in most districts. The places where young people congregate nowadays, in clubs and pubs for example, tend to draw customers from a variety of backgrounds; they remain part of the scene for only brief periods, then disperse to different futures.

Individualization does not mean that young people's family origins and achievements in secondary education have become unrelated to their future life chances. These old predictors remain in excellent working order (see Roberts and Parsell, 1992) and are proving resilient, but they currently operate in a variety of configurations. This means that young people themselves are less likely to be aware of all that they share in common with other members of any social category. They are more likely to feel personally responsible for their current circumstances and for building their own futures.

Third, young people's futures have become uncertain. It is more difficult than before for young people to know what types of adults they will become. This is partly a straightforward consequence of individualization. When large numbers from a cohort travelled into adulthood together, following a predictable life cycle, it was relatively easy for them to look ahead, at what had happened to earlier cohorts of young people like themselves, and glimpse their most likely futures. Uncertainty is also a consequence of the substantial numbers of young people who enter recently introduced educational courses and training schemes with no track records. Equally, it arises from the larger numbers on longer established routes, in higher education for example. There are now so many university students that their qualifications cannot unlock the same attractive career prospects for them all as rewarded earlier generations of graduates. Perhaps most basically, uncertainty is a consequence of the sheer pace of economic and social change which means that the adult roles that many of today's children will play are still unknown. This is a different situation to that which confronted young people in the 1960s and before. By age 19 the few who remained in full-time education were an academic elite who could be confident of entering commensurate employment. Others were completing apprenticeships and could expect to be skilled for life. Others knew that they were unlikely to obtain anything better than ordinary jobs. For young people today basing their self-concepts on what they hope to become is hazardous. Rebelling and dropping out from predictable futures have not become unfashionable so much as impossible.

Fourth, a corollary of their uncertain futures is that all the steps that young people can take have become risky. They cannot avoid risk taking. Higher education may lead to an excellent career but a university entrant today would be unwise to rely on this. Employer-based training may lead to a skilled job; if so the occupation may last for a long time. But there is simply no way of being sure. Personal relationships have become equally risky. Marriage may lead to lifetime domestic security or despair. Young people have lost much of their former security. Individuals have to take risks nowadays and the stakes are their own future lives. They have to travel towards adulthood without reliable maps, as if in private motor vehicles, albeit with differently powered engines since some have already accumulated advantages, rather than the public transport vehicles in which entire cohorts used to embark on the life course. Their new situation can appear threatening but some members of earlier generations felt stifled by their predictable futures. Uncertainty and risk taking can be liberating experiences. An indeterminate future can be more attractive than knowing the limitations of one's prospects. Today's young people would not all love to exchange places with their counterparts of the 1950s and 1960s, if they were able to do so. Needless to say, this is not among their options. Facing uncertain futures and taking risks are just normal parts of life for today's youth, and the majority appear to be fully prepared to take charge of their own lives. Their futures are not scripted for them, and they realize this. Many of the old 'rules' have simply disappeared. This forces young people to be responsible. They know that they need to be able to give their reasons for quitting or remaining in full-time education. Irrespective of whether they can explain their own behaviour convincingly, they know that they must face the consequences.

Fifth, young people's dependence on their families has been prolonged. Of course, families differ in their ability and willingness to discharge their prolonged responsibilities. The parents in the Dutch study of 120 young people (see Chapter 4) were nearly all extremely supportive. They did not try to exercise authority but advised and negotiated with their grown up children and tried to prepare them for 'choice biographies' (du Bois-Reymond *et al.*, 1995). Young people without families on whom they are able or wish to depend are at a heavy disadvantage. Family support is often crucial in enabling young people to complete their full-time studies and to make the transition into independent households (Coles, 1995; Jones 1995). Remaining in an intact family is a huge advantage (see Spruijt and de Goede, 1995). The young people who are still making quick transitions – leaving education and seeking employment as soon as they are legally able to do so, and becoming teenage parents – are mostly from heavily disadvantaged family and educational backgrounds (Kiernan, 1995).

Constant functions

It would be amazing if the above changes in young people's situations had not affected their leisure, given that leisure is so highly context dependent, and we shall see that there have indeed been changes but these have occurred alongside strong continuities. Leisure performs some functions for all age groups. Chapter 1 explained that the standard functions of leisure include allowing individuals to recuperate, to express themselves (to let off steam) and to acquire skills that may subsequently prove useful in other domains, and bonding participants into groups. Young people's leisure has always kept them apart, and continues to set them apart from adults because, being young and therefore having no deeply ingrained tastes, they are receptive to the latest fashions. There are additional functions that young people's leisure has performed in all modern (and some pre-modern) societies. It enables them to assert their independence from adults, typically with the blessing of their elders, and acts as a milieu in which they can learn to play sexualized roles and acquire the associated feelings and self-identities.

Children's play is not really leisure in so far as it is normally organized and supervised by adults. Children have to be taught to play, in approved ways, and to enjoy the experience. This teaching occurs daily in most families and schools. Children learn that there are correct times for play, and correct things to do in these times, and that playtimes should be enjoyed. They are given equipment to play with (toys), taught how to use them, and earn adult approval by playing correctly which includes expressing enjoyment. The ways in which children are taught to play vary by sex, social class and between ethnic groups. Child's (1981) study of the behaviour of 22–55 month old Birmingham children on play buses found that those from Asian families were the less active and the more respectful towards their play leaders. Children also learn to play in ways that are appropriate for their age and they earn adult approval, which is amplified in peer groups, by playing in ways which demonstrate that they are growing up.

Individuals begin to leave childhood behind when they start to play independently, usually with the approval of their parents and teachers who know that the young must learn to make their own ways into adulthood. Leisure is the arena in which most young people first learn to act independently. During youth there is a gradual trend away from spending leisure in settings such as youth clubs which are organized and supervised by adults towards spending time with groups of friends in unsupervised situations, then, later on, towards using commercial facilities (Hendry *et al.*, 1993). Young people assert independence simply by having their own friends with whom they develop their own tastes, and decide what to do with their time and money. Friends become key figures in young people's leisure. Simply being with their friends is extremely important to most young people. So are friends' opinions, but usually only on leisure matters such as which music to listen to, which hairstyle to adopt and which clothing to wear. In peer groups young people learn to deal with social equals without close supervision by anyone in authority, a skill that they will need in their future family and working lives. In developing such skills young people learn to claim their own space in streets and parks, within their homes (in their own rooms), and later on in settings such as sports centres, pubs and clubs. They learn to play the consumer role and to expect to be treated with the respect normally commanded by paying customers.

Establishing independence does not require young people to defy adults. Overtly defiant behaviour usually has some additional significance, connected with class-related attitudes towards authority for example (see Willis, 1977). Parents are usually delighted rather than offended when their children show that they are growing up. Activities such as smoking and drinking which are forbidden for children but permitted for adults usually have special attractions. Breaking the law is exciting for anyone but simply establishing independence does not require everyone to commit 'proving offences'. Nor is young people's need to assert their independence a sufficient explanation of the now widespread use of illegal drugs (see Measham *et al.*, 1994). Young people's willingness to experiment and try out new things is relevant, but the main appeal of leisure drugs among young people is based on the elementary fact that they make the users feel good.

The second function which makes young people's leisure different is that it is the arena where sexual roles are learnt and sexualized identities acquired. Societal expectations are not just transmitted but amplified in peer groups. Young people are under enormous pressure to prove to others, and to themselves, that they are sexually normal. This involves displaying the correct attitudes towards one's own and the other sex, and learning to treat members of each sex in the appropriate ways. Social pressures interact with biology to teach males and females not just to behave, but to feel masculine and feminine. Young people learn to display their emotions in gender appropriate ways. Girls learn to confide with same sexed friends. Boys normally learn to be independent and self-contained. It has been argued that male socio-emotional norms matched the kinds of conduct required in the more powerful occupations in the past, thereby disadvantaging women, whereas in present-day occupations which require more intricate interpersonal skills when dealing with colleagues and clients, femininity is more likely to prove advantageous (Halford and Savage, 1995; Nare, 1996).

Young people learn sexually appropriate conduct initially among same sexed peers, then in mixed sex crowds, then subsequently practise their skills in couple relationships. This is one of the processes that make youth cultures finely age graded. Young people are acutely sensitive to age distinctions. 'Mature' 16 year olds do not want to be confused with 14 year old 'kids'. Successive age groups claim space in different settings or have ways of keeping themselves apart in the same places. As they become sexualized young people develop and adopt their own norms for regulating sexual conduct. They decide whether it is OK to have sex in a casual relationship and judge each other accordingly. They also decide how to recognize if a relationship is steady. Relationships are usually so regarded if they are sexually exclusive and the partners communicate with each other (see du Bois-Reymond *et al.*, 1995). Within such relationships young people may feel that it is unnecessary to practise 'safe sex' though they learn rapidly that steady does not always mean permanent. Serial monogamy appears to have become the norm throughout modern societies, and this norm is initially established among young people.

Its special functions give young people's leisure a different significance even when they take part in the same leisure activities as adults such as going out for a drink, or staying in to watch television or to listen to music and converse. The crucial differences – the independence that young people find it necessary to assert and their sexual expressiveness – often alarm adult observers. Young people have always been regarded as a threat, liable to overturn normal patterns of family life and commit offences. Adults tend to regard themselves as potential victims but from the standpoint of the young it is they who are the vulnerable, harassed, victimized age group. Sheila Brown's (1995) questionnaire study among over 1000 11–16 year olds and interviews with 200 found that 30% of this age group could recall being followed by an adult on foot, and 18% recalled being followed by an adult in a car. Over one-fifth had been physically assaulted by other young people. The young people's main defence strategy was to stay with their own groups. Another such strategy in recent years has been to carry a weapon of some description. Most young people are telling the truth when they explain this behaviour – which most adults find offensive – as a defensive precaution.

The continuities in young people's leisure can easily beguile all age groups into believing that nothing has changed. Parents who are shocked by the rave culture and body piercing can recall that they themselves provoked outrage with their football hooliganism or punk styles in the 1970s and 1980s, or as mods or hippies earlier on. While expressing horror at today's youth, adults can comfort themselves in the belief that they understand present-day young people's behaviour and problems. Adults have always been able to respond to the latest youth scenes with such ambivalence. However, in the past, alongside the continuities, there have been major transformations in the significance of youth cultures. One such transformation occurred after the Second World War and the late 20th century may be witnessing an equally important shift.

Post-war youth cultures

Youth cultures pre-date 1945. Throughout the 19th and early 20th centuries young people in urban industrial Britain found space for themselves, usually on the streets, where they established their independence and built sexual reputations. But these youth cultures were local phenomena which occasionally acquired national reputations, usually for their toughness or criminality (see Davies, 1992). There was little commercial leisure targeted specifically at young people. Between the wars the dance halls catered for the young but the cinema, spectator sports and pubs had mainly adult customers. Young people were dependent on their families. Teenagers were paid boy and girl wages which, until they were 21 or became engaged, were usually handed over to their mothers in exchange for pocket money. Young people on the streets were generally considered 'at risk'. The boys were said to be at risk of becoming criminals while the girls risked pregnancy or, almost as disastrous, blighted reputations. Middle class families protected their children by keeping them at home, in schools and universities. Youth movements and clubs were intended for working class youth. Those not 'in contact', the 'unclubbables', were regarded as the high risk group. They tended to be from the most disadvantaged, often 'rough' rather than 'respectable', working class households.

After the Second World War a new kind of youth culture arrived. There were Teddy Boys and, later on, mods and rockers. These youth cultures were unexpected and unprecedented. At the time they were regarded as a threat to civilization. The new youth cultures were associated, in media discourse, with the rising rates of juvenile delinquency and teenage pregnancy. With hindsight, in practice from the 1960s onwards, it has been easy to see that these new youth cultures were products of full employment, narrowing differentials between young people's and adults' earnings, teenagers going 'on board' from the beginning of their working lives, and the affluent teenager becoming a market segment that was targeted by the suppliers of a range of leisure goods and services, especially music and clothing. When commercial options became available young people began to vacate the traditional youth clubs and movements (Albermarle Report, 1960). Young people's new ways of life, or at least their styles, gained unprecedented visibility. By virtue of how they dressed and their musical tastes young people could identify, and be identified with, flamboyant national youth scenes. In the 1950s it was mostly working class young people who left school at age 15 and earned 'good money' quickly who became involved in these youth cultures. In some respects these young people were simply an avant-garde group in the consumer culture and lifestyles that have subsequently spread to all sections of the population (see Chapter 7).

By the 1960s it was possible to see that the young people who became involved in these new cultural scenes were not threatening civilization as it was known. Sociological studies began to re-interpret the post-war youth cultures as processes of continuous socialization. The young people were not really dropping out. Actually they were making accelerated transitions to adulthood. Progression into adult employment accelerated and the mean age of marriages fell. Young people's new situations and their new youth cultures were enabling them to establish adult identities and play adult sexual roles at younger ages than their pre-war counterparts. The 'teenage rebels' of the 1950s and 1960s grew into the next generation of respectable parents, often retaining their teenage idols – Elvis, Cliff and Cilla, for example.

Another feature that became apparent as soon as researchers were able to stand back and identify the main patterns in post-war youth scenes was that they incorporated conventional gender and social class divisions. A generation war was not replacing earlier class struggles. The central actors in most of the new youth cultures were male. Girls had peripheral roles. Their 'lives of their own' were most likely to be based around 'bedroom cultures' (see Sharpe, 1977; Griffin, 1985). Participation in youth cultures was also class related. The new youth cultures of the 1950s and 1960s were mostly working class phenomena. Young people on working class trajectories made the most rapid transitions to adulthood and earned adult wages at the youngest ages. It was not just that the most active and youngest participants tended to be working class. It also became apparent that their youth styles often incorporated specifically working class values: about masculinity, the importance of solidarity between mates, being able to enjoy a good time and display disrespect towards authority (see Hall and Jefferson, 1976; Mungham and Pearson, 1976; Willis, 1977; Hebdige, 1979). Far from challenging, the new youth cultures were reflecting and helping to reproduce established gender and social class divisions. Through the styles that they developed or adopted, young people were expressing values, and sometimes addressing problems and contradictions, arising from their gender and class situations. Some of the studies that emphasized these features of the post-war youth cultures could be accused of ignoring the extent to which the participants were also addressing problems rooted in the process of growing up and overlooking the fact that the committed participants were not statistically representative of their social classes (see Smith, 1981). The crucial point remained that rather than a revolutionary threat the post-war youth cultures were socially conservative.

Another transformation?

Have there been any new developments of wider social significance, as opposed to just feeling significant to the young people concerned, in youth cultures since the 1970s? It is always important not to lose sight of the continuities but there have been a number of trends which suggest that youth cultures in the 1990s are not basically just the same as in the 1950s and 1960s.

First, a wider age group is involved in today's youth scenes. This is a result of transitions to adulthood being prolonged or, at any rate, becoming more variable in length. Today's 30-somethings are often still mingling with the young singles while some of their age peers are parents of teenage children. When he interviewed a sample of young adults who were visiting the night scene in Newcastle upon Tyne city centre, Robert Hollands (1995) found that his respondents were aged up to 31.

There has been speculation about whether youth cultures are disappearing (see Wallace and Kovatcheva, 1998). If so, youth unemployment will not be the reason. Young people without jobs face similar leisure problems to the unemployed in other age groups in Western countries. They have sub-normal rates of participation in virtually all types of recreation that cost money. However, most students and youth trainees, and many of the young unemployed, manage to participate in a wide range of leisure activities with the cash that they raise from parents, grants, loans and part-time jobs (see Roberts and Parsell, 1991). There are huge inequalities in young people's incomes and spending levels but their spending patterns prove that it is possible to participate in most youth scenes at different levels of expense. Nights out, holidays and new outfits can cost a lot or can be managed much more cheaply. If youth cultures disappear this will not be a consequence of young people's poverty. It is more likely to be a consequence of pre-teen children being introduced to youth fashions in dress and becoming an important market for pop music, individuals staying young into their 30s, and the spread of popular consumer cultures into adult age groups.

Nowadays the main age division in musical tastes, above which the appeal of pop and rock dip, is not between teenagers and adults but above and below the 40-somethings (Longhurst, 1996). Unlike in the 1950s, it is no longer only, or even mainly young people who purchase leisure wear and recorded music. Consumer values and their associated culture have ceased to be as age specific as formerly. However, it seems unlikely that youth cultures will become indistinguishable from other popular consumer cultures if only because young people remain particularly susceptible to the latest trends, their leisure still performs some age-specific functions, and (see below) becoming a parent continues to make a substantial difference to day-to-day and week-to-week leisure.

A second post-1970s trend has been that gender and social class divisions within youth cultures have become less clear cut. This is not to say that either type of difference has disappeared. Despite unisex fashions and the gay villages in some cities, most young (and older) people succeed in looking unmistakably male or female, and heterosexual masculinity and femininity are expected if not demanded in most informal social settings. One change is simply that most youth cultures are not as male dominated as in earlier times. Young women have broken out from the bedroom culture. The recent changes in their leisure behaviour are certain to be related to the fact that young women are making more headway in education (out-performing males at all levels) and in the labour market, and gaining effective control over their own fertility. Young women are now as likely as young men to use indoor sports facilities (Department for Education, 1995).

They have been claiming space in other public places also: city centres, wine bars and throughout the club scenes.

There has been a similar blurring of social class divisions. Since the 1960s popular culture has been adopted by young people on middle class trajectories in secondary schools' academic streams and in higher education. This has not driven out high culture; it is more a case of young (and older) people now feeling able to enjoy both classics and pop. There is more intermingling of the social classes in comprehensive secondary schools and in post-compulsory education. This, along with the less certain futures of all young people, has made them harder to 'classify' by researchers, and by one another. An outcome is that there are no longer any clear social class differences in the kinds of music that young people listen to, the fashions that they wear, or the leisure places they go to. Robert Hollands (1995) found that Newcastle city centre night life was equally popular among university students and local young people. Those from middle class home backgrounds, who are educationally successful, still tend to participate in more leisure activities, but the 'more' is of the same kinds of things in which working class youth are also involved. Middle class youth, and males more so than females, are more likely to become committed participants and to play leadership roles in all kinds of leisure. For example, they are the more likely to be involved in running sports clubs and playing popular music, but the activities themselves are not restricted to, or even ostensibly associated with any sex or social class (see Roberts and Parsell, 1994).

A third trend has been the splintering of youth cultures. Needless to say, there are still many things that most young people do. The majority go on nights out and consume alcohol, listen to popular music, watch television and play sport (see Roberts, 1996). However, there are no longer any particular musical genres or fashions that can claim to be the dominant style. There is a rapid turnover in Top 40 numbers and artists. None seem able to exert the appeal of the Beatles, Elvis and Cliff. New technology is part of the explanation. The music production and distribution industries have more players. There are more radio stations, all trying to appeal to specific taste publics. But technology is not the reason why it is now equally fashionable to have short-cropped or long hair, or why there is no 'uniform' worn as widely as blue denim in the 1960s and 1970s. Young people's tastes do not map neatly on to either social class, gender or geographical divisions (see Roberts and Parsell, 1994). There is simply more variety within all social groups which will somehow be related to the broader processes of individualization. Young people use leisure to develop and express their individuality but can only do this via their specific sub-cultural affiliations.

A fourth trend has been towards young people's sub-cultures acting as a base for proto-communities (Willis, 1990) or 'new tribes' (Maffesoli, 1994) rather than expressing membership of pre-existent groups. The groups of young people (and adults) who become fans of specific sports teams, and who attend raves and similar scenes where their preferred types of music are played, can experience intense camaraderie. Much of the appeal of these occasions is that they are incredibly social. Everyone is friendly. Individuals find that they are accepted. None of this is completely new. The change over time has been that the young people who play together nowadays have rarely grown up together in the same districts and attended the same local schools. And their participation in the 'new tribes' is usually temporary. Participants are always drifting off to other scenes. Sometimes the sub-cultures simply disappear.

> Club cultures are taste cultures. Club crowds generally congregate on the basis of their shared taste in music. . . . Taking part in club cultures builds, in turn, further affinities. . . . Clubs and raves, therefore, house ad hoc communities with fluid boundaries which may come together and dissolve within a single summer or endure for a few years.
>
> (Thornton, 1995, p. 3)

Yet being part of these scenes is extremely important to young people. Those studied by Robert Hollands (1995) who were part of Newcastle's downtown night life were going to the city centre 2.7 times a week on average and spending £16–18 per occasion, which amounted to 38% of their total incomes. They nearly all said that they would feel unacceptably restricted if they were unable to experience these nights out.

It is plausible to argue, though difficult to prove, that through their subcultural affiliations today's young people do not so much express as acquire social identities and self-concepts. If education, occupations and gender no longer confer the clear identities that they offered in the past, if today's young people cannot be sure about the kinds of adults they will become, if they cannot know what their occupational or social class destinations will be, they may need to become immersed in specific sub-cultures in order to define who they are. Young people could be the vanguard age group in this respect. In the 1950s and 1960s the new youth scenes signalled the spread of consumer cultures. Maybe today's young people are similar pioneers, using lifestyles which incorporate their preferred uses of leisure to create their most significant, identity conferring, social positions. Some say that this is the future of leisure, which will become more widespread as the current cohorts of young people carry their styles into adulthood. This suggests that leisure is now acquiring a role that extends beyond the economic, psychological, social and political grounds for treating it seriously which were itemized in Chapter 1.

These questions about young people's leisure – whether it is really becoming the base for their most significant social identities and self-concepts, and whether this role of leisure must be expected to become more widespread – are taken up again in Chapter 8. Until now the leisure of older age groups has remained more conventional.

Adult Leisure

Adulthood has been affected by the destandardization of the life course. The impact may have been less dramatic than among young people up to now, but while young people's futures have become extremely uncertain, adults' actual lives have become less stable in many cases, and less secure for the majority. Adulthood today is at greater risk of disruption due to occupational restructuring and the dissolution of domestic partnerships. There is also more variety in the pace and order in which people make the transitions normally involved in becoming adult: marrying and having children, living independently from their parents, and supporting themselves financially through their own or their partners' employment. Individuals need not make any of these transitions in order to gain social recognition as adults. However, despite the tendency for these transitions to occur at later ages, for more 30-somethings to remain part of the singles scenes, and for more women in this age group to remain childless, the vast majority of the population are still making all these transitions at some point or another. As regards the leisure implications, the crucial transitions are to living with an opposite sexed partner and the arrival of children, and the effects seem little changed from when there was a more reliable life cycle.

The squeeze

Estes and Wilenski (1978) coined the term 'life cycle squeeze' to describe the consequences of marriage and parenthood. Their vocabulary is still the best available. The squeeze occurs as constraints increase: mortgages to pay off, households to run, and more dependents to support. These constraints make new demands on time (see Table 5.1) and money. An outcome is that adults' leisure is vulnerable, especially women's, since up to now men have been more skilled, or perhaps just more able to protect their own pastimes (see Chapter 4). Nevertheless, living with an opposite sexed partner, and even more so the arrival of children, lead to an all-round reduction in out-of-home leisure for both women and men. Spare time, such as remains, tends to become home based and television viewing increases. When adults go out this tends to be as a family, on holidays and day trips for example. Once again, this is especially true for women. Younger people's leisure-based lifestyles, and their alleged identities, seem unlikely to survive this squeeze. The squeeze is not slackening. The total working time (paid and unpaid) of parents of dependent children is not only longer than in any other section of the population but has been lengthening rather than diminishing mainly because more mothers are in paid employment without any compensatory reduction in the time that fathers spend in their paid jobs. Total workloads now tend to rise more steeply with the birth of children than was the case in the past (see Zuzanek *et al.*, 1998).

There is more variation than before in the ages at which the squeeze begins. Nowadays it can be at any point from teenage to beyond 40 though the longer child bearing is postponed the more unlikely it becomes. Voluntary infertility seems less likely to occur from a firm decision early on never to have any children than from a series of decisions not to have them 'now' until a point is reached when aspirant parents realize that reproduction will be difficult and it is too late to do anything about the problem, or the adults have reached career and income levels where they feel that they have too much to lose (Hobcraft and Kiernan, 1995). Nevertheless, some adults begin their careers as parents when aged 35-plus, a time of life when some of their age peers are relinquishing their main child rearing responsibilities. Children today remain dependent for longer so it is by no means exceptional for these responsibilities to extend beyond age 60 by when the adults may have become substantially responsible for the care of their own ageing parents – an implication of older people living longer.

Family instability

The current instability of adults' lives does not alleviate, but intensifies the squeeze. A mother's own leisure may be more vulnerable than her male partner's but lone parenthood is no solution. Lone parents, who are nearly always women, tend to have extremely home-based leisure lives. They are more likely than other women to ask, 'What leisure?' (Streather, 1979; Hardey, 1990). Their domesticity is enforced rather than being the mothers' preference in most cases; it is a product of lacking the time and money to spend on their own pleasures. When adults embark on serial partnerships this is likely to intensify and prolong the squeeze. Absent parents (usually fathers) can find their time and money drained by having to contribute to the support of two households and their desire to retain contact with their children. When second partnerships yield more children the squeeze is extended. All the evidence points to the leisure benefits of the conventional intact family. Young people benefit from the support that such families can provide during their prolonged transitions and this is evident in the young adults' physical and mental health, and relatively low risks of unemployment (see Spruijt and de Goede, 1995). Adults' leisure, men's and women's, also benefits more often than not, from an intact relationship.

Time and money pressures

Washing machines and other items of domestic technology should have reduced the time squeeze, and higher rates of employment among mothers should have alleviated the financial squeeze if all other things had remained equal. In practice, however, financial pressures tend to be exchanged for time pressures or vice versa. As Chapter 2 explained, expected standards of home and child care have risen. Moreover, in certain respects present-day societies have become child-unfriendly. Parents are more reluctant than in the past to simply send their children out to play. Streets and parks are often considered dangerous. Parents often feel that it is necessary to escort or transport their children to and from school, and to and from leisure activities. Needless to say, not all parents are willing and able to be so dedicated. Sometimes an outcome is that children are expected to amuse themselves with televisions and other items of leisure equipment. Parents who are willing and able to take their children out may find the occasions stressful. They are likely to find that young children are unwelcome in many adult leisure settings. McKeever's (1993) study of catering establishments in the Oxford area found that few made any special provisions for young children; their parents often complained of being made to feel like second class citizens.

Social class differences

There are social class differences in the severity of the squeeze. The middle classes' advantages are based more on money than time. As observed when discussing work–leisure relationships, money can alleviate if not solve most leisure problems. Better-off parents are better placed to spend on child care, to preserve their own nights out and maintain existing leisure interests. They are also more able to take their children out for leisure, coping with the stresses. This has been illustrated vividly in Fishwick and Hayes' (1989) study of 401 Illinois adults. The arrival of children was associated with increased sports activity among the wealthier respondents; children were an additional reason for going to sports centres. In contrast, poorer parents' sports participation declined with the presence of children. Allan and Crowe (1991) have pointed out that home centredness need not be at the expense of wider sociability. Acute privatism is more likely to be an outcome of constraint than preference (see also Devine, 1992). Children are nearly always an additional constraint on their parents' leisure, but more so in working class than in middle class households, and most of all when the parent is single.

The most representative life stage

Adult leisure inevitably displays most of the general features of leisure in present-day societies. It is normal leisure writ large in the amounts of time spent at home, with the family, and watching television. The normal gender and social class differences become more pronounced than in earlier life stages. Holidays are a major item of leisure spending, and the adults' own out-of-home leisure, when they have the opportunity to devote occasions to themselves, is most likely to involve going out to eat and/or drink.

Despite the current level of marital breakdowns, one of the main pleasures that most married partners experience is the other's company. Oriel Sullivan's (1996) research, which involved 380 couples keeping time budget diaries for a 5-day period, found that activities were most enjoyed when undertaken jointly. This was particularly true of leisure activities, and most of all when the leisure involved socializing with other people. The couples in this research appeared to be trying to coordinate their schedules so as to increase the amounts of time that they spent together. Even so, the proportions of their activities done simultaneously were only slightly higher than for pseudo-couples (males and females paired at random). Pseudo-couples do many things at the same times because, despite the destandardization that is in process, most people still work during week days, watch television in the evenings, and so on. In so far as individuals deviate from such normal schedules, as a result of working odd hours for example, it seems to be difficult for couples to coordinate their activities and off-set the disruption of their togetherness even when they wish to do so.

Leisure at home

The compensation for the squeeze is supposed to be that adults obtain homes of their own in which they can savour each other's and their children's company. Some of this is just hopeful idealism. We know that homes are sometimes places of stress, violence and abuse. However, we have seen that most couples do in fact value each other's company.

In 1982 Sue Glyptis and Deborah Chambers drew attention to how little is known about leisure in the home. This remains the case despite occasional forays such as Sullivan's and despite the long-standing knowledge that the home is where most leisure time is spent. Part of the explanation for the gaps in our knowledge is that home life is private, impossible for outsiders to observe unobtrusively. Researchers have to rely on what their informants show and tell them. However, when they conducted a study of leisure and the home among an East Midlands sample, Glyptis and Chambers found that nearly all their subjects were extremely cooperative (Glyptis *et al.*, 1987). Home-based leisure remains rather mysterious not so much because people are unwilling to say what they do but because unless sexual and other intimate details are revealed the picture is unremarkable. Home life appears to be composed of chores, television and small talk. Can it really be as mundane as the reporting makes it appear? When we stand back, try to detach ourselves and adopt an external view of our home lives, the banality can be difficult to explain.

Glyptis and her colleagues conducted a time–space analysis of their subjects' daily lives. They investigated what their subjects did, with whom and where, measuring not just how much of their time was spent at home but how the space in the dwellings was used. Their evidence, which can be made to sound either unremarkable or surprising, showed that day-to-day activity tended to be concentrated within a communal living area in most households. The residents were not making the fullest possible use of all the space that they had at their disposal. There were few examples of cellular living despite the electricity, central heating, studies and garage workshops that some of the dwellings contained. Spending leisure alone at home seems to be common only when no one else is in. At least part of the explanation will be that family members really do enjoy each other's company. But another part of the explanation could be that a principal function of much home-based leisure is straightforward recuperation. Or perhaps we are slaves to tradition and addicted to television. The evidence from Glyptis and Chamber's study is difficult to square with Gordon Cherry's (1984) portrait of homes bulging with leisure equipment that the occupants lack both the time and the space to use. Most homes appear to contain a great deal of under-used space and leisure equipment.

Patterns of family life in Britain appear little different from in Eastern Europe where the typical dwelling has just two or three rooms, shared by two or three generations, and all the rooms are used for both living and sleeping (see Roberts and Jung, 1995). The hours that the typical adult devotes to television do not suggest that time scarcity prevents the audio equipment, computers and sports gear in Western homes from being used more frequently. It could be that the owners derive satisfaction from leisure equipment simply being available if and when they wish to use it. It could be that there are advantages in upstairs rooms often being used for sleeping only so that household members do not have to go to bed simultaneously or interrupt each other's leisure when they do so. Maybe these returns are considered sufficient to justify our investments in our homes. Or do the types of dwellings that we buy, and the ways in which we equip them, reflect our aspirations and the images that we prefer to present to visitors rather than how we actually live?

Maybe the Chinese are more realistic. Freysinger and Chen (1993) report that the family is not a preferred leisure milieu in China. They attribute this to Confucianism, the family being regarded as the primary cell of society and properly ruled by authority and obedience rather than free expression and affection. In Britain, as in China, real leisure appears to mean going out for young people and adults alike. Young people willingly sacrifice this leisure, maybe having learnt its limitations, then have little option but to lead more home-centred lives due to a combination of financial necessity and their obligations to dependents.

Leisure in Later Life

Change and continuity

The recent literature on growing older in later life (the ageing process begins at birth) is as brimful as youth research with references to the life course being destandardized and deinstitutionalized, and chronological milestones being removed (Laczko and Phillipson, 1992; Young and Schuller, 1991; Kohli *et al.*, 1992;). Whether the changes in later life justify the discourse is debatable. The change in older people's circumstances that has prompted the references is that more men have been terminating their employment careers, or having their careers terminated for them, at younger ages than in the past. More men have been taking voluntary retirement, or have been compulsorily retired early, or have been losing their jobs then unable to regain work, before age 65. This milestone, or millstone, has not so much been removed as changed in significance. Sixty-five is still the age at which state retirement pensions become available for men and it will have the same significance for women when sex equality is attained as European law requires. Most occupational pension schemes have always operated on the assumption that people will pay contributions, or have these paid on their behalf, for 40 years in order to qualify for full pensions, and 65 remains the normal age beyond which there is no contractual obligation on an employer to retain an employee. The change is that in the past most men knew that they would have to work, or be able to continue working, depending on how they felt about their jobs, up to age 65. At this age they knew that their main careers would end. Nowadays most men must wonder how close to 65 they will be when their working lives are terminated. Some fear being cast on the scrapheap. Others wonder whether they will be lucky enough to be offered enhanced early retirement packages. In the past some men resented having to work up to age 65. Others resented being compulsorily retired merely because they had grown 65 years old. However, up to the 1970s, in the decades of full employment, it was not uncommon for men to continue working either full-time or part-time, in their existing jobs or in other occupations, well beyond age 65. This is much rarer today. The flexibility is now up to age 65. This milestone has not disappeared but changed in significance. Men's reactions differ; as we shall see, some have welcomed the change while others have felt victimized.

There have been other changes affecting the population's experiences in later life. More people are now retiring with occupational in addition to state pensions. More women are accumulating pension entitlements in their own right and there has not been a decline in the proportion of women working beyond age 55. People are now living longer; old age lasts longer on average than in the past. This means that many younger old people, in their 60s for example, are helping to support older relatives. The costs of supporting the ageing population are being transferred from the state welfare budget, the National Health Service and local authority social services departments on to the 'customers' and, to some extent, their younger relatives.

However, the process of ageing in later life has always been lengthy. It begins, in a sociological sense, when people begin to disengage from what have been their main lifetime roles. One such disengagement is from child rearing. This usually occurs gradually rather than abruptly as children grow older and more independent and leave the parental home one by one, but at some stage there has normally been an 'empty nest'. Another disengagement is from employment. This has sometimes occurred abruptly but it has often been gradual via part-time, temporary and voluntary work and this remains the case today though the transitions tend to occur earlier in men's lives. Other life events happen inevitably at some stage – bereavement and the loss of physical and mental capacities.

Growing old involves a series of transitions and changes. There has never been just one big abrupt step. Compared with young people's transitions, those in later life normally occur over a longer time period, involve relinquishing rather than taking on new roles, and the individuals concerned have rather less control over their life events. Whereas young people have some say over when to terminate their education, when to marry and whether to become parents, older persons have less control over when their children leave home, and when their own physical and mental capacities decline. Some have no choice over when to terminate their main lifetime employment. And even when they have some say over 'when' there is an inevitability about the events happening at some time or another.

Disengagement and engagement

A theme in the literature on leisure in later life is that disengagement from the major roles of adulthood can be an opportunity to engage in other things. This is the Third Age of Golden Years perspective which emphasizes that disengagement from family and work obligations can open opportunities for a leisure renaissance in which people do things that they really value.

These accounts contain more advocacy than description. They are about what ageing could be like rather than what it currently is like for most people. We are living and staying healthy for longer. Many of us relinquish family responsibilities and retire from employment with decades of active life ahead. With mortgages paid off and no dependents, some households have unprecedented cash for leisure spending on holidays, theatres, meals out and so forth. It is not difficult to find people acting in precisely the way that the engagement theory prescribes. The Woopies (well off older persons) have become a major segment in several leisure markets, especially holidays. Thousands spend the winter months in warm climates. This age group provides most of the customers on ocean cruises. At any holiday resort, coastal or countryside attraction they can be seen alighting from or boarding their coaches and filling the hotels throughout the year. Some become leisure connoisseurs, experts on what the holiday trade, stately homes, art galleries and museums have to offer. Max Kaplan, a leading American writer on leisure, has been among the advocates of the engagement view of ageing. His main book on this topic (Kaplan, 1979) was prompted by his own official retirement. In it Kaplan draws attention to how ageing, and leisure, offer opportunities for commitment, company, service, aesthetic and tourist experiences.

Many obstacles have been identified to making the leisure renaissance an opportunity for everyone. Ill health and poverty restrict many, but other obstacles might be removed more easily. One is ageism which views the elderly as lacking in capabilities and even appetites. This societal perspective may even influence the ways in which ageing persons view themselves. Another obstacle is said to be the low regard that many people still have for leisure. However, Betsy Wearing (1995) has argued that leisure offers alternative discourses with which ageing people can prevent their devaluation. A leisure perspective, she argues, can highlight the many things that ageing people are able to do, and she offers examples of older persons using leisure to enrich their lives and to demonstrate to others that older people should be targets for envy not pity, and that the Third Age is something that everyone can anticipate eagerly. Robert Stebbins (1998) argues that serious leisure, when people have such interests, can be the basis for fulfilment during retirement.

Leisure trends in later life

The aggregate picture is well known, does not appear to have changed over time, and bears little resemblance to the engagement perspective. As people grow older the general tendency is for them to reduce their leisure activities. Specifically, the range is reduced, especially in out-of-home recreation. But as previously explained, ageing takes a long time and older people rarely make major changes in their leisure from year to year. Their leisure is more stable than in any other age group. Young people are far more likely both to adopt new activities and to drop former interests (see Mihalik *et al.*, 1989). Older people tend to have very fixed routines. It is tempting, but maybe unfair, to describe them as being stuck in ruts.

In the early 1980s Jonathan Long and Erica Wimbush studied a sample of men in Scotland shortly before and just after they retired from employment (Long and Wimbush, 1985). This study recorded very few examples of retirement being seized as an opportunity to develop new leisure interests.

Nor were there many examples of sudden disengagement. Overall there was a slight decline in leisure activity. In some ways retirement was having similar effects to unemployment; people were having to adjust to lower incomes and the loss of work-based social relationships. Very few of the men in this study had received or made careful preparations for retirement. Most simply carried on and adjusted as best they could; the momentum of their lives seemed to carry them along. Their leisure resembled the unemployed's in that they spent most of their time at home and, compared with the general population and their own lives prior to retirement, they watched more television, did more reading, gardening and listening to the radio.

The overall tendency among the ageing is for life events such as retirement from employment and declining health to be accompanied by a reduction in leisure activity. However, the aggregate is made up of many different experiences. It is known that, at some points in later life, some people increase their involvement in certain kinds of leisure, taking holidays for instance. It is also known that the elderly are over-represented in some forms of out-of-home leisure, using parks, going to bingo, churches and playing bowls for example. Later life lasts a long time and covers a huge variety of circumstances. The population experiencing later life varies widely in age, physical and mental capacities. There are wide differences in health within all age groups, and there are equally stark contrasts in social and economic circumstances. In later life some people retain strong family connections. In some cases these act as a support system though it is often the ageing who are the supporters of grown up children and grandchildren, and more elderly relatives. Inequalities of wealth and income are wider among the elderly than among people of working age (Abrams, 1980). Some older people retain the use of their cars whereas others find their mobility restricted by their lack of private transport, poor public transport, declining physical capacities and lack of income.

The stability of older people's routines is actually a short-term phenomenon. Very few change their habits abruptly in any year but, as already stressed, later life can last a long time, longer than adulthood (the period when they are established in their main lifetime roles) for some people. In Germany, Walter Tokarski's (1991) research among a sample initially of 222 older men and women, declining to 52 over 15 years, recorded a great deal of lifestyle change. People were changing their ways of life as their economic, domestic and health circumstances changed. Over the 15 year period major change was normal, indeed inevitable in many cases as people grew older, lost partners, saw their savings exhausted, and so on.

Early retirement

There have been several studies of early retirees, mainly men. There has been the same gender imbalance in research into leisure in later life as in other fields. Retirement from full-time, long-term employment careers, a momentous transition for more men than women, has attracted more attention than changes in domestic roles including bereavement which is more likely to affect women since men tend to die at younger ages. The studies of early retirees are nonetheless interesting because of their conflicting evidence.

On the one hand there have been studies such as Young and Schuller's (1991) among mainly working class men in London who lost their jobs well ahead of the normal retirement age. These men are portrayed as in a limbo; too old to have any real hope of regaining decent jobs but too young to consider themselves, or to be regarded by others as properly retired. Most of these men were bitter and disillusioned by the tattered and undignified endings to their working lives. On the other hand there are studies such as Elizabeth McGoldrick's (1983). She encountered little bitterness in her questionnaire survey of around 1800, and interviews with over 100 early retirees. Most of McGoldrick's sample had volunteered to retire early and only 6.5% expressed regret at having done so. These retirees were healthier and wealthier than the retired population in general. Few had taken early retirement on account of failing health. Nor had many been inhibited by serious financial worries; most had quit with attractive severance packages. Some had volunteered for early retirement in order to escape from negative aspects of their jobs – the pressure, the other people, or the constant travel, or whatever else was irking the individuals concerned. However, most had been at least equally if not more motivated by the prospect of new lives, sometimes involving new jobs, but often based on hobbies or being able to spend more time with friends and families.

Jackson and Taylor's (1994) study of 175 unemployed men, all aged over 51, who were followed up from 1982 to 1983 offers some pointers into what separates the winners from the losers when working lives are shortened. They found that individuals with the financial and other resources to build lives and identities outside the labour market gained psychological benefits, including a feeling of being in greater control of their own lives, by considering themselves retired. Those who could not, or had not opted for this solution to their unemployment were mostly struggling unsuccessfully to regain work.

Leisure may be hopeless as a general solution to unemployment (see Chapter 3) but this will not prevent it from being part of the solution for some members of specific sections of the population. People who end their working lives ahead of, or at the normal retirement age, can retain the status and identities associated with their former careers. They become retired professors, plumbers or whatever. If individuals can be guaranteed pensions that secure their standards of living, and if they have leisure interests to which they wish to devote more time, these may well perform some of the functions normally associated with employment – structuring time, providing activity, goals and social contact. Needless to say, when early or normal age retirement occurs on the basis of who volunteers it is likely that those who take the step will be those who are able to benefit by doing so.

Leisure and later life satisfaction

All the relevant studies show that high levels of leisure activity in later life are related to life satisfaction as self-assessed in questionnaire studies (see Box 5.1). Leisure is related to expressed life satisfaction in all age groups (see Chapter 1) but this relationship is especially strong among the elderly. It is never easy to prove whether a statistical relationship is spurious or causal and, if the latter, in which direction. It is likely that people who suffer chronic health problems will be generally less active and satisfied with life simply on account of their ailments. However, it is no less likely, and consistent with all the evidence, that older people with leisure interests that keep them busy and provide social contact will find their lives more satisfying than those who spend nearly all their time at home, watching television for 50 or more hours per week. Since elderly people are less likely than other age groups to have paid jobs and heavy family responsibilities, which nevertheless supply satisfying experiences, leisure differences can be expected to discriminate their general life satisfaction more powerfully than in other sections of the population.

Box 5.1. Leisure and the quality of later life.

Strong relationships between leisure participation and satisfaction with later life have been recorded in all the countries where such studies have been conducted.

In a study of 360 retired people in Puerto Rico, Nelson Melendez (1992) found that leisure activity was more strongly related to satisfaction with life than any other variables including health and income.

A USA study of 1649 over-55 year olds found that their leisure activities were satisfying needs normally met by paid work in younger age groups (Tinsley *et al.*, 1987).

Leisure was supplying companionship, opportunities for self-expression and social recognition.

It was also supplying some of the sample with a sense of power since their leisure interests gave them a chance to run things.

Among the over-55s leisure could confer these benefits without individuals being niggled by the feeling that they really ought to be in paid employment.

Questions about why people differ so considerably in their levels of leisure activity in later life, and why some rather than others have sufficiently strong leisure interests to tempt them into early retirement, are of practical as well as academic importance. Satisfactory health and income are important for well-being in later life. These advantages will facilitate leisure activity but they are not the whole explanation of why some of the elderly are far more leisure active than others.

Leisure Careers

Most research on leisure and the life course has compared the leisure patterns and activities of different age groups, and explored how these change during youth, family formation, and up to and then in later life. But there is another complementary way in which life course trends in leisure can be analysed, which is to examine changes and continuities over time in the behaviour of the same individuals. Such research is not easy. Panels can be studied longitudinally as in Walter Tokarski's (1991) study (see above) but this type of research takes a long time before it yields longitudinal data. Fortunately with the more structured leisure activities it has been shown that most people have sufficient powers of recall to reconstruct their leisure biographies (see Hedges, 1986). At any rate, their accounts look realistic in terms of the sex and social class differences that emerge, and the overall participation rates that are indicated. This kind of research can produce quick results and it has greatly enlarged our ability to study long-term leisure careers. However, it shares one major limitation with panel studies, namely, that the social contexts may have changed considerably since people currently in later life were young, so their leisure biographies in youth, and the factors that shaped them, could be poor guides to what is happening today.

Massive continuities

Nevertheless, the findings from research that has traced individuals' leisure biographies are startling in the strong continuities that are revealed. In the USA Scott and Willits (1989) restudied in later life a sample that had been surveyed initially when young, 37 years earlier, when information was collected about their uses of leisure. This research showed that the people who were involved in voluntary organizations, who had wide circles of friends, and those with intellectual and artistic interests in later life were nearly always the same individuals who had been doing these things when they were young. The exceptional people who were taking up new leisure activities in later life usually turned out to be following a lifelong pattern. Likewise most of those who were reducing their leisure participation following retirement had been narrowing their leisure interests throughout adulthood (see also McGuire *et al.*, 1987).

The best predictor of any individual's future uses of leisure is that same person's past behaviour. This is a far better predictor than the individual's current circumstances – type of occupation, employed or unemployed, for example. An individual's past leisure behaviour usually proves to be an excellent predictor from day to day, week to week and, indeed, across the life span. Most of us use the relative freedom of leisure to stick to our routines. In our leisure most of us are basically conservative and become more so as we grow older. We tend to stick to familiar routines and do tried, tested and trusted things. Taking risks at play may be safe compared with the potential losses at work and in family life but very little leisure behaviour is experimental.

Early leisure socialization

Personality factors can account for some of the continuities over time. Sociable people will tend to make friends throughout their lives. However, early leisure socialization appears to be extremely influential. This has been demonstrated in the arts and sport. In every country high culture is a minority interest. Everywhere nearly all the members of this minority become involved in the arts early in their lives, usually in their families, but sometimes in education. Individuals who attend classical concerts or visit art galleries regularly when they are young are very likely to continue doing these things throughout their entire lives, whereas the rest of the population is most unlikely to take to high culture during adulthood. The crucial role of early socialization into the arts and the importance of the family as a socializing milieu, led the coordinators of the international study which collated the information, to express pessimism about the prospects of enlarging substantially the audience for high culture (Hantrais and Kamphorst, 1987).

Sport is not exactly the same. Most children play sport fairly regularly but in most cases this does not lead to lifelong sports careers. Two factors appear crucial to whether people will remain sports active into adulthood. The first is the richness of their early sports socialization, not the sheer amount so much as the number of different sports that individuals learn to play. Young males whose skills and enthusiasm are confined to football are unlikely to have long-term sports careers. The second crucial factor is whether individuals remain in sport during their transitions from education into the labour market, marriage and parenthood. Individuals who stay in sport through these life events are likely to remain sports active for the remainder of their adult lives. In contrast, individuals who drop out before their mid-20s are unlikely to return to sport on a permanent basis. If their sports careers are restarted they run high risks of lapsing again. It proves much easier to keep the committed in sport than to convert 'couch potatoes' (see Roberts and Brodie, 1992).

The extremely low rates of participation in physically active recreation within the present-day older age groups are not due solely to them having dropped out of sport during their adult lives but are also due to the poverty of their childhood sports socialization and the limited opportunities that were available for them to continuing playing sport after leaving school (see Boothby *et al.*, 1981). This is a case where there have been major changes since people currently approaching retirement age were young. School children are now being introduced to a much wider range of sports than in the past, and indoor sports centres have been opened throughout the country so young people have become much more likely to continue playing after completing their full-time education (see Roberts, 1996). This means that nothing further need change in order for sports participation in the older age groups to rise: if all other things remain equal the current cohorts of young adults will carry their higher propensity to participate into later life.

Implications for leisure providers

Promoters of particular forms of leisure appear well advised when they target the young, but the contrasts between the arts (where rich childhood socialization alone appears to be a sufficient foundation for long-term interest) and sport (where breadth is necessary and drop-out during young adulthood remains common) suggest that the details of a success strategy will depend on the type of leisure. And since it is impossible for every adult to be a frequent participant in every kind of leisure, the programmes of different promoters may partly cancel out each other's efforts. However, some adults succeed in doing more than others with similar leisure resources. Many older persons lead limited leisure lives partly because they have few interests on which to draw. The findings from research into long-term leisure careers suggest that this situation can be changed, but the bad news is that achieving major changes in the population's leisure habits is more likely to take decades than months or years. The best recipe for spreading any use of leisure, and maintaining participation at a high level, appears to be, first, to ensure that children are introduced to the skills and the satisfactions that are available, then secondly to ensure that they remain active for long enough to become locked in not just by routine but through organizational memberships and social commitments, and a desire to continue to benefit from what they know are reliable sources of satisfying experiences (see Roberts *et al.*, 1991a). This same recipe seems likely to work equally well in religion and politics.

The reasons for the higher socio-economic strata's general tendency to do more include their early leisure socialization. Middle class homes where the parents have wide-ranging leisure interests are rich learning environments. Higher education can be particularly valuable in extending young adults' leisure interests and skills, and maintaining their involvement into their 20s. The unemployed have low rates of participation in out-of-home leisure not just because of their immediate circumstances but also because the sections of the workforce at the greatest risk of unemployment (the least qualified and skilled) have the lowest levels of leisure activity. Retired professionals are active in more forms of leisure than retired manual workers partly because the former were already doing more prior to retirement.

Up to now childhood and youth have been the main life stages for leisure capital formation. This is not to suggest that leisure learning ceases once individuals become adult. Skills and interests may continue to be acquired throughout life. In adulthood, as opposed to childhood, the learning is likely to be deliberate and conscious (see Erickson, 1996). The key point is that adults do not learn haphazardly but incrementally, on the basis of existing skills and interests. This is why childhood and youth leisure socialization are crucial. Most people base the rest of their leisure lives on interests, which may subsequently be built upon, to which they were initially introduced when young. This would change if the life course was totally destandardized. Leisure learning would then become equally likely at any age. In practice, however, destandardization is not being pushed this far. Marriage and parenthood continue to trigger a leisure squeeze. In later life the general tendency is still for people to narrow down their existing leisure interests and activities. Wider social and economic trends have not shattered any of the main leisure patterns: the gender differences, the tendency for the economically privileged to do more of most things, or childhood and youth being the life stages where the foundations are laid for long-term leisure careers. As with financial arrangements, the best time to begin preparing for leisure in later life appears to be as young as possible.

The evidence from long-term sports and arts careers may foster pessimism about the prospect of policy initiatives or anything else achieving quick changes in the population's leisure patterns, but the long term prospects are far more encouraging. In the future it is likely that leisure will supply many people's most dependable threads throughout the life course: social relationships, activities and satisfactions on which they know they can rely in societies where so many other things have ceased to be reliable. People's leisure is affected by life events but leisure also affects individuals' ability to cope. We have seen (see Chapter 3) that if leisure activities are maintained during unemployment this is good for the well-being of those concerned, and the unemployed are far more likely to persist with existing uses of leisure than to start anything from scratch. Likewise later life satisfaction depends largely on the leisure skills and interests that individuals carry with them from earlier life stages. The leisure careers evidence points to the counterproductivity of seeking quick results from policy interventions. It shows that secure and lasting effects need a long-term strategy. Whether we wish to use the findings is a different matter but leisure research tells us how to widen the entire population's leisure repertoire and, in the long term, boost overall levels of leisure activity in all social categories.

Chapter 6

Experiencing Leisure

Criticisms of Residual Concepts

The preceding chapters have repeatedly hit problems in applying the leisure concept adopted at the outset of this book to people's actual lives. These problems were anticipated in Chapter 1. Pristine concepts are always liable to grate on encountering a murkier real world. This applies with definitions of education, the family, religion, social class and work as well as leisure. Some degree of jarring when concepts meet reality has to be tolerated. But despite all that has been achieved in researching the growth of leisure, and in exploring how its distribution and uses are related to people's ages, sex and occupations or lack of any, critics allege that the amount of jarring when leisure is defined as 'time left over' is simply too great to be tolerated.

The non-employed

First, there are problems in identifying the leisure of the non-employed. Some have commitments that can be treated as the equivalents of paid work: pupils and students who attend schools and colleges full-time for example. But few of the unemployed and the retired have commitments or obligations that are close equivalents of paid work. In trying to study their leisure, questions arise such as whether they lead entire lives of leisure or have no leisure at all. Both propositions look ludicrous. And the non-employed are not a small, insignificant minority. Less than half the populations in Western countries are in paid employment, and the non-employed sections are growing. There are more unemployed than in the 1950s and 1960s, and the numbers who are retired are not only higher than in the recent past but are expected to grow steadily throughout the early decades of the 21st century. Roughly three-quarters of all free time is 'enjoyed' by people who are not in full-time jobs (Martin and Mason, 1998). Such trends add weight to appeals for a more realistic leisure concept.

Women

Second, it has always been difficult to apply the residual concept to women's lives irrespective of whether they have paid jobs. This is due not just to the quantity, but to the character of women's domestic obligations. In women's lives these obligations are not normally time bounded. Women who are the principal, albeit unpaid carers of dependent children, and nowadays increasingly the elderly, have never had fixed hours of work. Hence the case for 'feminist' research methods and concepts that identify leisure as a feature of tasks, relationships or experiences rather than as blocks of time.

Blurred boundaries

Third, the residual concept has never been totally consistent with the real lives of male employees. Just as some women have experienced going out to their paid jobs as an escape from domestic drudgery and tedium, so some, possibly most, men have obtained leisure-like experiences at and from their paid work. This becomes apparent as soon as investigators ask what people lose when they become unemployed. In most cases they miss of the companionship of work colleagues and the interest, challenge and sense of accomplishment that they found in their occupations.

Trends

Fourth, it is argued that current trends are fast outdating the residual concept of leisure and diminishing the merits that it once possessed. Work time is being destandardized. Alongside this the notion of there being set times for leisure interests and activities could also be rendered obsolete. Women still do most of the housework but men, albeit slowly, are becoming more involved. The life course is changing in ways that mean that there are no longer specific leisure patterns and practices that can be linked to particular age groups. There is a pervasive process of individualization which requires individuals to negotiate their own paths through life. In the case of leisure this means accumulating personal 'capital' from various sources, not just specifically leisure settings, during successive life stages, and finding ways to use this capital to advantage in situations subsequently encountered during the course of careers in family life, paid employment and leisure.

Rob Lynch and Tony Veal (1996) distinguish modern leisure from several earlier varieties (see Chapter 1) and also from postmodern leisure which is said to be spreading. According to Lynch and Veal, in modern times work and leisure were clearly demarcated; leisure time was otherwise residual time which could be used for rest, entertainment or self-development, and was serviced by specialized leisure industries. They believe that in present-day postmodern conditions the old boundaries between work and leisure have become fractured and fluid, that (for some) there has been an explosion of opportunities to spend money on 'stylized' goods, services and activities whose main value lies in their culturally assigned meanings, and that leisure thereby becomes a component of broader lifestyles whose significance extends well beyond the now traditional functions of modern leisure.

It can be argued that some leisure pursuits, which used to be very occasional and therefore out of the ordinary, have now become so common as to break down the everyday/out-of-the-ordinary distinction. According to Rojek and Urry (1997) this applies to tourism. Many of us travel, and encounter visitors in our home areas, so frequently, that all this may be regarded as mundane rather than an escape from routine. This is just one example of a blurring of the divisions which once made leisure a more easily recognized part of life.

The mounting problems with residual concepts are exposed, some argue, by the growth of uses of time and money that cannot be clearly assigned to leisure or anything else. Time spent in education, especially post-compulsory education, is one example. Part-time students, whose numbers have risen, might be regarded as choosing to devote some of their free time to education, but many of the students would argue that their education is at least partly vocational and that they regard their time spent studying as work. Full-time students will usually regard their education in this way to some extent, but they may also agree that the intrinsic satisfactions and general attractions of student life are among their reasons for being at college. Time spent shopping and on personal care have also risen. In the past leisure researchers usually treated shopping as a chore and self-maintenance as among life's necessities. Nowadays it can be argued that much window shopping, and much of the time that we spend attending to our bodies, ought to be classified as leisure. The sums that households spend on food (for home consumption as well as when eating out), on transport, housing, clothes and telephones usually include a leisure element. None of these are minor items. But it is difficult, if not impossible, to decide how much of the costs of a motor vehicle, for example, should be regarded as leisure spending rather than as a domestic or work requirement. All these examples, it can be argued, illustrate the breakdown of 'modern' boundaries and the break-up of residual leisure.

Some leisure researchers have always seen advantages in defining leisure not as a type of time but as a kind of experience, and they are now able to claim that history is on their side. They can argue that the rise of modernity was accompanied by a dominant form of time organization in which most things had proper times and places, whereas in the era that we are entering these links are more tenuous. Even some who doubt whether contemporary societies are really experiencing a postmodern transformation still welcome postmodernist thinking for rescuing social thought from the illusions of modernity under whose influence, it is claimed, we imagined that our 'grand narratives' could explain, or even change, a real world that was always far less orderly (Rojek, 1995).

It is worth noting that no one recommends defining leisure as a type of activity. It has always been just too obvious that any activity might be leisure or work. Gardening and playing sport are examples. Some people are paid to watch films and television to monitor the output. Simply observing behaviour has never been sufficient to establish whether the activity should be classified as paid work, unpaid work or leisure. The debate among leisure scholars has always been whether an activity should be regarded as leisure depending on when it occurs or whether the actor's experience should be the arbiter.

Leisure as Experience

Decentring leisure

Decentring, a little used word outside the social theorists who are part of the 'postmodern turn', has been introduced into leisure studies by Chris Rojek in his book *Decentring Leisure* (1995). Writers have different ways of describing what decentring means. Rojek himself explains that something is decentred when we realize that 'meaning and reality are structured by language. What we take to be truthful and objective is merely an effect of language . . . (Leisure) is part of the representational and symbolic machinery that we use in order to negotiate daily life' (p. 131). This is said to be a radical view in so far as

> most contributions to political economy assume that real needs can be identified and that effective policies can be formulated to address them. Decentring violates these assumptions. It treats needs and policies as links in a metonymic chain in which meaning is permanently unstable . . . In the light of this critique leisure studies is reduced to the mere play of signs and symbols
>
> (pp. 131–132).

In other words, if, when we develop theories about leisure, we believe that we are making statements about a real world out there, we are gravely mistaken. In fact what we take to be knowledge is simply a product of our use of language.

David Chaney gives an entirely different meaning to decentring leisure: 'a move away from more public, collective ways of participating in cultural occasions towards more private, personal modes of participation' (Chaney, 1996, p. 113). Nicos Mouzelis (1995) offers yet another definition of decentring. He says that to decentre involves putting at the centre of an analysis 'not the individual subject, but social practices that are, in a manner of speaking, disconnected from the actors who have generated them' (p. 45). This may seem to contradict Rojek's argument that to decentre is to recognize that practices that might appear to exist 'out there' owe this appearance to the language that we use. In other words, what appears to be objective and true, including the meanings of these words themselves, is embedded in the beholders' vocabulary. The aspect of this that Mouzelis stresses is that language is always a collective thing. It develops, but exists before any individual uses it. Meanings are already part of the language and in this sense they do exist 'out there', in the language and its meanings which are not created by any particular individuals or groups.

Everyone who uses a language is likely to accept its view of the world as objective truth. This applies equally to the discourses of lay people and scientists. Decentring (in Rojek's sense) makes everything relative. The pedestal of objectivity is kicked away. Truth becomes merely a discursive convention. Once scholars realize that there are always many possible ways in which things can be constituted discursively they are expected to cease imagining that their own discourses are superior 'grand narratives' which offer the correct interpretations of events. They are expected to abandon this 'modernist' project and . . . well, they may conclude that there is little else for them to do but to reflect discursively on their postmodern predicament. Their only option, so it might appear from their own theory, is to become a tribe of consenting adults who admit converts to an inner-circle upon proving that they accept the articles of faith, can use the language and respect the house rules of discourse.

None of these ideas are new but they have come to the fore in late-20th century social thought because a critical mass of postmodern thinkers, represented in most of the social sciences, have taken the radical implications seriously. Some believe that their own frame of mind is shared by increasing numbers of lay groups, in which case postmodernity is held to exist beyond social theory 'out there' in society. However, symbolic interactionists and sociologists influenced by phenomenology have a long record of not only recognizing but insisting that actors' definitions of situations are crucial to how they respond. Mainstream sociology has long been sensitive to there being many possible readings of any situation, and that phenomena such as social classes, deviance, poverty and leisure do not exist as ready made structures for citizens to access and researchers to study but are constituted and reproduced only because people believe in them and apply these labels to particular times, groups, events, activities and experiences.

Social theorists who believe that 'reality is in the mind of the beholder' have normally sought an escape from the predicament in which they might appear to be locked by their own theory. They have usually tried to claim a privileged position, maybe achieved through reflexivity, taking account of the relative status of truth, building this into their own rules of discourse, and thereby gaining knowledge of other groups' realities which amounts to more than a construction grounded in the theorists' own linguistic conventions. Having 'escaped' they have proceeded just like other 'modern' scientists except, possibly, that they have accepted their need to operate with a heightened sense of permanent and pervasive contradiction (Ang, 1996). If a privileged position is deemed impossible there would appear to be nothing left for theorists to do except to work discursively within their own theories, attributing truth and status to each other's work, but without any pretence of generating objective knowledge about an external world. However, anyone who proceeds to write about leisure 'out there' in society is making an escape attempt; having decentred the subject, the first question that such enquirers are likely to ask is, 'How do people recognize it in their own everyday lives?' Investigators cease treating phenomena as things that can be observed independently of lay people's perceptions and experiences, and realize that a thorough understanding must be through these experiences. Many leisure researchers have long believed that the starting point for their enquiries should be lay people's experiences and definitions. Decentring is a new, and rather tortuous, route to a well-established position.

Everyday concepts

Many researchers have asked members of the public what leisure means to them. This is matter of some interest, however leisure is defined by the investigator. In order to explain the leisure behaviour of specific groups it may be considered useful if not absolutely essential to establish what they think leisure is. All the relevant enquiries have produced the same kinds of answers (see Stockdale, 1986; Samdahl, 1988; Mobily, 1989; Kleiber *et al.*, 1993). Some groups, the unemployed and housewives for example, often query whether the term leisure has any relevance to their own lives, but this implies an understanding of what leisure is like for other people. Leisure has proved a meaningful concept for all groups within modern societies, and their everyday understandings highlight three features.

First, it is common for leisure to be distinguished from work but, second, when lay people appear to opt for a residual definition they usually state or imply that leisure is different because work is disagreeable or has to be done. Their everyday understandings associate leisure with choice, lack of constraint, being able to express oneself, and doing things voluntarily. The word 'freedom' often crops up in these contexts. Third, people also refer to leisure as being pleasurable or enjoyable and sometimes relaxing, and to these experiences being immediate or intrinsic.

All sections of the populations in all modern societies appear to share very similar understandings of leisure. They may have different views on whether they have enough, and the quality of what they have, but there appear to be no fundamental disputes about what leisure is.

Derived definitions

Leisure researchers who opt for experiential definitions can claim to be respecting and responding to their subjects' ideas, and can recommend their concepts on these grounds. Christine Howe and Ann Rancourt (1990) have proposed a definition of leisure as 'a self-enhancing affective state'. John Neulinger's (1990) definition is 'self-motivated conduct that contains its own reward'. The particular forms of words are less important than how this entire group of definitions tries to respond to ordinary people's conceptions of leisure as self-determining and fulfilling.

Optimal experience

Many writers who operate with experiential concepts have been attracted by the work of Mihaly Csikszentmihalyi (1990, 1993), an American social psychologist who has conducted a series of investigations to explore what people mean when they claim that something is enjoyable. Csikszentmihalyi's proposition is that enjoyment depends on the balance between the challenge that a situation or task poses and an individual's skill in the relevant area (see Table 6.1). When an individual's skill is overwhelmed by the immensity of a challenge, Csikszentmihalyi argues that anxiety will be experienced. If their skills are such that a challenge can be met almost effortlessly, he argues that people will soon become bored. When both an individual's skills and the challenge are modest people are said to become apathetic. Peak enjoyment is said to arise from challenges

Table 6.1. Mihaly Csikszentmihalyi: Flow.

	Skill	
Challenge	Low	High
High	*Anxiety*	*Flow*
Low	*Apathy*	*Boredom*

in areas where individuals are highly skilled, and when their capabilities are stretched to the utmost. In such situations Csikszentmihalyi argues that people will become wholly absorbed and 'lost' in the activities, that they will become unaware of other issues and incidents in their environments, and the passage of time. Csikszentmihalyi has adopted the term 'flow' to describe this state.

Some leisure researchers, though not Csikszentmihalyi himself, have sought to associate flow with leisure. When such connections are made it is not being suggested that when at leisure people experience constant flow. However, it can be claimed that when this 'optimal experience' is not available in other areas of people's lives, they are likely to seek and achieve it through leisure activities. People may experience flow in sporting contests, where optimal experience depends on meeting an opponent of equal ability, or they can become thoroughly absorbed in dancing, playing music, listening to music, reading or whatever. It is often suggested that optimal experience is most likely to be achieved through leisure. For example, Rojek (1995, p. 182) argues that, 'As paid work becomes routinized and mechanical leisure becomes the axis for the development of creativity and personal enrichment'. Even if people are not gaining optimal experience through leisure it can be argued that they could be, and perhaps should be, at least seeking it.

Whatever view is taken on the merits and relevance of Csikszentmihalyi's flow concept, experiential approaches can be commended for raising questions about what people are seeking and deriving from leisure. Eric Dunning (1996) has criticized the conventional sociology of leisure (and everything else) for failing to pay proper attention to the sensual and emotional aspects of life. In the study of leisure he believes that this neglect is inexcusable since, according to Dunning, one of the principal roles of leisure is to counterbalance the emotional staleness of 'civilized' societies. The figurational sociology which Dunning advocates requires behaviour to be analysed in its total context, but only having paid full regard to the minutiae of the immediate setting and the behaviour itself. As soon as we approach leisure in this way, Dunning believes that we cannot fail to recognize that it is all about expressing emotions – in fact most forms of play are designed to be emotion raising – and often requires sociability and movement also.

Experiential approaches start from the 'inside', with what people at leisure are actually doing, and their immediate contexts, motivations, satisfactions and frustrations. Researchers find themselves asking what people mean when they claim to enjoy something, what it means to have fun and what makes people laugh. Detailed studies have shown that laughter can be provoked by exposing the private thoughts that people are likely to harbour but keep to themselves, when an alternative view of everyday situations is suddenly 'switched on', or when normal hierarchies and relationships are abruptly overturned (see Mulkay, 1988; Podilchak, 1991; Mulkay and Howe, 1994). Explanations of fun turn out to be disappointingly sombre.

Serious leisure

This is a term coined by Robert Stebbins (1992), a Canadian sociologist, and is another product of getting into leisure from the inside. Stebbins defines serious leisure as 'the systematic pursuit of an amateur, hobbyist or volunteer activity that is sufficiently substantial and interesting for the participant to find a career there in the acquisition of its special skills and knowledge' (Stebbins, 1992, p. 3). Instead of dividing uses of leisure into tourism, sport, television viewing and so on it becomes possible to distinguish the serious from the casual and to recognize that serious leisure may link experiences in ostensibly different domains such as listening to radio programmes about, and taking holidays that are devoted to, an interest. Stebbins has drawn attention to the fact that some people become so involved in their leisure interests that they become as skilled and knowledgeable as professionals in astronomy, sport, drama and so on. He argues that serious leisure enables people to derive a long-term sense of accomplishment, which may, provided people have such leisure, operate as a functional alternative to work during retirement or in the event of unemployment (see Stebbins, 1998). Enquiries based on residual social structural definitions can be criticized for being unlikely to recognize such possibilities.

Apart from Stebbins's own case studies, up to now there has been little systematic research into serious leisure. However, Stanley Parker and a team of colleagues have studied the serious leisure of a group of Australians. There were only 30 subjects in this study, but they were able to supply information about 222 of their leisure activities. Of these, only 17% was judged to be serious and another 18% was judged partly serious. The arts, handicrafts and volunteering were the leisure activities which were most likely to be serious. Sport was rarely more than partly serious, while watching television, reading and socializing were nearly always casual. Middle class males aged over 50 were more likely to report serious leisure interests than other respondents (Parker *et al.*, 1993).

Leisure and self-development

There is not just one residual, social structural definition of modern leisure, but several. Investigators who operate with residual concepts make slightly different choices of the boundary and contextual features to highlight. The same applies with experiential concepts where there is a major split between those who use experiential definitions alongside or within, and those who propose their concepts as alternatives to, residual definitions.

John Kelly, an American sociologist, is associated with the former type of experiential concept. He defines leisure not simply as time that is otherwise uncommitted and which is therefore free for individuals to use in any way that they wish, but as a particular kind of way in which this time can be used. Kelly (1983, 1987, 1994) has successively described his approach to leisure as existential, symbolic interactionist, and critical constructionist, but his preoccupation has always been with how leisure is experienced. He argues that modern societies create space, leisure space, in which people are not only able to express, but create roles and identities for themselves. Leisure is said to provide 'freedom to be'. Kelly argues that our capacity not merely to act parts, but actually to create roles and identities for ourselves, is uniquely human. Human behaviour is not wholly programmed by biology. Nor are we just social products who do no more than enact the roles into which we are socialized and rewarded for compliance and punished for deviance. Along with many other contemporary sociologists, Kelly emphasizes our ability to be 'reflexive', which means that, far from just mirroring our biological natures and social contexts, we are able to stand back, take an 'outside' view of ourselves and the roles that are mapped for us, develop ideas about how both might be changed, and act upon these ideas. According to Kelly, this has become possible to a greater extent in leisure than in any other sphere of life. Leisure can provide this freedom because, according to Kelly, leisure is 'beyond necessity'; our societies have no reason to restrict us as severely in leisure as in other areas of life, and therefore actors can experiment and let themselves go without dreading the consequences of failure.

Primitive playfulness

Chris Rojek's experiential definition of leisure differs from Kelly's in several ways including the absence of any social structural accompaniment. In all his work Rojek has criticized the 'social formalism' which treats leisure as a product of a particular kind of society, and which seeks explanations of its uses in the actors' other social roles (Rojek, 1985). Rojek regards such treatments of leisure as part of a failed modernist project which has sought to explain everything rationally and has believed that, once explained, affairs might be brought under rational control. The 'donatory' approach to leisure policy – the idea that good leisure can be delivered to the people by policy makers and programme operators – is another alleged failure of the modernist project. Rojek has tried to persuade leisure researchers to dispense with these illusions. He is sceptical of claims that we are becoming a different kind of postmodern society but is attracted to postmodernist thought in so far as it recovers 'what the illusions of modernity have concealed' (Rojek, 1995, p. 192).

Rojek believes that what is now called leisure springs from a pre-social human impulse to seek pleasure. He believes that these impulses are beyond wholly rational control either by individuals themselves or by their societies, and that the impulses are always liable to erupt, often in surprising ways.

Therefore, leisure experiences are likely to occur chaotically while we are at work, at school, at political or religious gatherings and, of course, in specifically leisure behaviour. The job of leisure research, in Rojek's view, is to identify the outcomes of human pleasure seeking, and to study the ways in which various interest groups endeavour to control other people's pleasure seeking, to contain it within rules – spatial, temporal and behavioural boundaries – in accordance with the prescribers' particular moral precepts or commercial interests. However, it follows from its elementary nature that pleasure seeking can never be totally controlled.

Ning Wang (1996) has offered a similar conceptualization of leisure. She suggests that in all societies it has been possible to recognize two broad types of behaviour. First, there has been conduct governed by social structures and rules whose character has varied by time and place. She points out that all societies have possessed an 'order' with which members have been expected, and constrained to comply. Second, whatever the social, economic and political order, Ning Wang argues that 'Eros' has always found expression: people have done some things spontaneously, simply for diversion, amusement or fun. She argues that part of the modernist project has been the confinement of Eros within particular parts of life that are generally described as leisure. This particular division, she recognizes, has been a characteristic of modern societies. In contrast, she claims, Eros itself, the source of what we call leisure, is not specifically modern.

Universal relevance

An advantage claimed for experiential concepts is that they escape from the notion that only modern societies have leisure. As Ning Wang points out, all peoples seem to have found ways to do things for pleasure, for fun. This was recognized in Chapter 1. Maybe people in pre-modern times had no words corresponding to 'leisure' but it is unnecessary to be able to spell or even say leisure in order to enjoy it. Hultsman (1995) has argued that our preoccupation with definitions is specifically Western. In other civilizations people have been satisfied that something simply 'is'. Leisure may be organized differently in modern societies, channelled through specifically modern time structures and activity patterns, but all cultures have had games and pastimes, have provided some space for people to be self-determining, and to enlarge their abilities and self-concepts. In most pre-modern societies the equivalents of modern leisure appear to have been less commercial and less individualistic. Play was more likely to have expressed communal solidarity than each individual's distinctiveness (de Grazia, 1992; Somnez *et al.*, 1993). Experiential concepts permit comparisons which often suggest that contemporary leisure is relatively impoverished. Whatever view is taken on this, the possibility of the comparisons can be applauded.

Many writers have noted affinities between modern leisure aspirations and a Greek ideal which was expressed clearly by Aristotle (see, for example, Hemingway, 1988). Aristotle approved of the manner in which the freemen of Athens had the opportunity to develop all their talents: their aesthetic tastes, physical prowess, intellectual faculties and political skills. The Aristotelian ideal was the fully developed, well-rounded individual who could move gracefully, act, throw, run, read, debate and participate in civic affairs. Marx appeared to share this ideal with his vision of a society in which people might work in the morning and hunt or write poetry during the rest of the day (see Rojek, 1984). The fact that such ideals echo through history can be offered as proof that it is only the form, not the inner substance of our leisure today, which is distinctively modern.

Policy relevance

Experiential concepts can be inspirational. The 'litmus tests' for definitions of leisure (and everything else) in Chapter 1 included conveying the main features, and experiential concepts certainly highlight some of leisure's positive qualities more successfully than residual, social structural definitions. Experiential concepts draw attention to what leisure does for people, and to what it might achieve.

The analyses to which residual concepts lead are relatively bland. People are portrayed as using their time to watch television, go on holidays, drink and suchlike. These descriptions are unlikely to ignite enthusiasm. Experiential concepts focus not only on what people do with their leisure but also upon how they are driven and affected. Maybe people are not in constant 'flow'. Perhaps they feel fully self-determined only on exceptional occasions. Maybe we are not all using our leisure to become everything of which we are capable. Experiential concepts allow for this, but also offer a view of what leisure could be. There is a messianic tinge to some treatises based on experiential concepts. For example, John Neulinger (1990) claims that the self-expression and fulfilment that were possible only for freemen in ancient Athens are now within the reach of everyone.

Bringing such a state within everyone's reach can be made into a goal of leisure policy. In comparison, residual concepts look sterile to anyone seeking policy implications. Experiential concepts enable writers to envisage the kinds of leisure opportunities that would be needed for everyone to derive all the benefits that leisure can offer.

Extending the research agenda

Residual social structural concepts can be criticized for limiting the research agenda, and being unable to engage with all aspects of the role of leisure even in the modern societies from which the definitions are said to derive. For example, there have always been important cultural dimensions to modern leisure. People do not just behave, but do so in ways that are meaningful to themselves and others. Leisure goods and services have meanings. Their purveyors are often keen to associate their wares with extremely attractive meanings. Experiential concepts appear the better equipped to explore the dynamics of consumer behaviour and consumer culture. These concepts also endow people with creative capacities, and thereby permit examination not only of what people do but how they can make roles, social positions and identities for themselves. These are in fact major topics in the agendas of subsequent chapters which advance beyond conventional enquiries into the effects of income levels, hours of work, gender and age roles on the leisure of different sections of the population.

Limitations of Experiential Concepts

Experiential definitions have attractive features. This book does not attempt to conceal these assets. But against these must be set a list of problems.

Operability

First, experiential concepts are difficult to incorporate in quantitative research covering large representative samples. This is especially the case when leisure is defined solely as a type of experience with no stipulation that the experience should occur in any particular part of people's lives. Working faithfully with this kind of definition would require researchers to study the whole of their subjects' lives, as in time budget enquiries, but in this case they would need to identify the occasions when individuals felt self-determined, when their behaviour was intrinsically rewarded, when they felt that they were having fun, in flow, or whatever the definitive experience.

Obtaining this information is not impossible. Csikszentmihalyi has developed a method which is called experience sampling. Subjects have to be persuaded to carry bleepers and are equipped with questionnaires. The bleepers are programmed to alert the subjects at random (except that there is usually a guarantee that the process will not operate during night-time). Whenever he or she is bleeped the subject has to complete a short questionnaire usually giving details about what he or she is doing, where and with whom, sometimes along with answers to a set of questions about his or her feelings.

The method works, but up to now it has been used only with small samples. The costs of large-scale research are prohibitive. This does not invalidate experiential concepts, but, as explained in Chapter 1, stipulative definitions are never right or wrong but simply more or less useful. Being operable is one of the 'litmus tests' and some experiential concepts stumble at this hurdle. A consequence is that, despite their appeal, the definitions have not become the base for systematic and cumulative research. Even researchers who advocate a feminist concept of leisure have often resorted to quantitative methods, and a 'masculine' residual concept, in their own large-scale studies of women (see Green *et al.*, 1990). Works that employ experiential concepts tend to be speculative, anecdotal and assertive, but never really convincing. Rojek believes, and is persuasive in arguing, that leisure expresses a basic human impulse to seek pleasure, but he is unable even to explain how this could be proved, let alone to offer the evidence. John Kelly presents the creation of roles and identities as definitive features of leisure but has no evidence to show that people are or feel more creative in their leisure than in other areas of their lives.

Coherence

Another problem is that experiential concepts, when otherwise unqualified, do not dovetail tidily with other concepts that their authors use. If leisure is a type of experience that can occur anywhere and at any time, it will include pieces of life that are commonly described as work, education, family interaction and so forth. Purely experiential definitions are inconsistent with leisure's everyday popular meanings in so far as these distinguish it from work that has to be done. The scope for confusion is immense. How can we judge which is the more satisfying if an experience can be both work and leisure? Claims for leisure that are intended to refer to experiences that might occur at any time and in any place are always liable to be misunderstood as referring to people's experiences in their leisure time.

Definition or evidence?

The coherence problem is avoided when leisure is defined as a kind of experience that is obtained or sought in leisure time, but such conceptualizations can be accused of settling, by definition, matters which should really be left to the evidence. The number one argument for not stipulating the experience when defining leisure is that discovering whether people feel self-determined, enhanced, intrinsically motivated, or whether they have created new roles and identities for themselves, should surely be treated as matters for investigation rather than resolved in advance. It appears mischievous to exclude by definition the possibility that leisure might be boring. It is difficult to imagine any good reasons for deciding in advance of any investigation that leisure is the sphere where people will be most able, or most likely to achieve self-enhancing experiences or states of mind. Ellery Hamilton-Smith (1992) has pointed out that people are likely to seek optimal experience, whatever this might be, in all areas of their lives rather than solely through leisure.

Csikszentmihamlyi (1993) himself has used experience sampling to investigate where people are most likely to experience 'flow'. His evidence shows that people are more likely to experience flow at work than during their leisure, and that the state of mind characteristic of most leisure is apathy. Judith Brook (1993), also, has found that in general people find their work more challenging than their leisure, though it is during leisure that they are most likely to feel 'in control'. Rejecting experiential definitions is not the same as rejecting proposals to investigate leisure experience. There is no dispute that leisure experience is important and residual concepts do not obstruct its investigation. The point at issue is whether it is useful to pre-define leisure as any particular kind of experience.

Experiential concepts may sometimes tell us more about the writers' hopes than about other people's leisure. John Neulinger's (1990) definition allows him to envisage the entire populations of economically advanced societies achieving the fulfilment that people have sought, unsuccessfully in most cases, throughout history. With humanity's potential fulfilled, we are invited to contemplate the peoples of the world relaxing in universal contentment and harmony. Deprivation and conflict would be banished from this civilization with leisure. This kind of vision may be an attractive professional ideology for the operators of leisure services. Leisure scholars can give themselves an escape route from their academic networks and become prophets and preachers. Social science easily becomes a casualty.

The actual evidence suggests that people seek and obtain not just one but many kinds of experience through their leisure – fun, company, relaxation, exhilaration and so on. They experience and plan for highs and lows, peaks and troughs. Sometimes people want excitement whereas at other times they wish to 'switch off'. Research can best investigate the variety of leisure experience by avoiding pre-judgements or, to put it more bluntly, prejudicial terms of reference.

Experience in Leisure

There is an overwhelming case for studying people's experiences in leisure, and the following chapters dwell on these subjective aspects. However, in doing this they build upon, rather than offer an alternative body of knowledge, to all that has been learnt from the more 'conventional' enquiries that supplied most of the evidence in previous chapters. The next chapters delve into meanings, but without specifying in advance which particular kinds of experiences are to be regarded as leisurely. Deciding this in advance is a sure route to colourful but thoroughly misleading conclusions. Perhaps more attention should be paid to ordinary people's experiences and their immediate contexts than hitherto in the conventional study of leisure, but the balance of advantage is decisively towards doing this within, rather than at the expense of the conventional leisure concept. A residual social structural definition merits retention for several crucial reasons.

Modern leisure is distinctive

Modern societies do in fact create a specific kind of leisure time, enable people to spend money for specifically leisure purposes, and have goods and services which are targeted specifically at this market segment, in ways that set them apart from pre-modern societies. And in this respect present-day societies remain basically modern. This is not to reject the desirability or the possibility of comparisons. Indeed, the very act of defining leisure as specifically modern makes a clear contrast. It remains possible to inquire whether the games, play and overall quality of life in societies with modern leisure are superior or inferior to what was available in earlier times. However, it is unlikely that the relationships between occupations and leisure, and unemployment and leisure for example, that are found in modern Western societies, and which have proved broadly similar in all such countries, will have any relevance in pre-modern situations where there has been no comparable division of life into work and leisure. Modern work–leisure relationships could not have preceded modernity, and a modern definition of leisure ensures that this is recognized.

A dominant pattern of time organization

It is true that residual concepts rarely map perfectly onto anyone's life, and are very difficult to apply in many cases. But the residual pattern can still be dominant even if the majority of a population is not in employment. This is because household schedules tend to be organized around the needs of their employed members, which thereby affect all residents. In a similar way, leisure events tend to be scheduled so that the prime wares and widest choices are available during the periods that are leisure for most workers. People without jobs cannot ignore this dominant rhythm. Even if they try to ignore it, the likelihood is that they will still be affected by it. If some of the retired and the unemployed, for example, are completely divorced from this rhythm they will have difficulty in experiencing normal leisure. This is likely to be their fundamental leisure problem which can be identified only by using a residual concept.

The persistence of modernity

This is one of this book's overall arguments: a post-industrial society, certainly, but postmodern? The previous chapters have recognized that there are major changes in process: the destandardization of working time, full-time employees in some Western countries working longer rather than reducing their hours, and greater variety in the circumstances and behaviour of men and women in all age groups. If these trends continued indefinitely all the older divisions and structures would be obliterated but, as yet at any rate, we are nowhere near such a point. Most of the research-based evidence pointing to the persistence of modernity is in the following two chapters, but before proceeding to this there are theoretical, a priori, grounds for scepticism towards claims about the eclipse of modernity. The contrasts between the modern and postmodern that are drawn by those who believe that the former is being superseded by the latter do not suggest that the transformation will be as thorough as the earlier change from traditional to modern society.

Table 6.2 lists some of the keywords associated with traditional societies in the left column, modern industrial societies in the centre, and postmodern or post-industrial societies on the right. Some of the changes said to be currently in process are broadly equivalent to the earlier shift from tradition to modernity. Work shifting from agriculture into manufacturing then into service sectors, and jobs from agrarian to blue-collar to white-collar, and the organization of life from local, to national to global levels, are of this order. In contrast, leisure cannot replace work in quite the same way. Nor can consumption replace production in the same way that, earlier on, science displaced religion, and life was divided into work and

Table 6.2. Traditional, modern and postmodern societies.

Pre-industrial	Industrial	Post-industrial
	Fordist	Post-Fordist
Traditional	Modern	Post-modern
Agriculture	Manufacturing	Services
Peasants	Blue-collar	White-collar
Self-sufficiency	Production	Consumption
Way of life	Work	Leisure
Religion, tradition	Science	Imagination
Community	Structure	Culture
	Groups	Individuals
Locality	Nation state	Globalization

leisure thereby superseding the manner in which groups used to produce for their own needs. People can have leisure and consume only if work is done producing things, and production continues to be based on the application of science. The implication is that conventional 'modern' leisure concepts, and the findings based upon them, are not becoming totally obsolete. Proceeding as if leisure and consumption were transcending former structures, and as if culture was escaping from all its earlier moorings, leads to grossly exaggerated claims, fears and hopes for leisure. Rojek (1993) himself has acknowledged in his discussion of present-day tourism that all 'escape attempts' ultimately fail; people are obliged to return to modern realities.

Recentring leisure

Calling for leisure and other phenomena to be decentred re-opens a long-running debate in sociology in which investigators are invited to choose between what, in reality, are perfectly compatible positions, such as recognizing the significance of structure and agency in social life. There could be no social structures unless people were sustaining and recreating them by behaving in appropriate, meaningful ways. But the aggregated actions of people who share understandings of social class, religion and leisure give all these phenomena the capacity to confront every individual as if they were external realities. There would be no such thing as society without individuals, but this does not mean that there are only individuals and no society.

Whatever the original source of people's leisure motivations – nature or nurture – sociology's principal interest must be in how these motivations are socially shaped and controlled. And this shaping and controlling is not solely by the 'rules of pleasure' that various groups seek to prescribe but also by the distribution and organization of time and income, forms of leisure provision, and the particular opportunities and constraints that arise from gender and age roles, types of employment and unemployment, and social class positions more generally. People's own ideas about leisure must certainly be taken into account in explaining their leisure behaviour, but these ideas do not exhaust what leisure really is. The extent to which leisure behaviour is amenable to rational, scientific explanation is not a matter that can be settled a priori but only by the success or failure of the efforts, and the results up to now have been encouraging. Leisure behaviour has been shown to be related systematically to many other things. Paying attention to people's authentic leisure experiences adds a further dimension, leads to more comprehensive answers to older questions, and opens up new ones, but without in any way discrediting more conventional findings or their foundations in residual social structural concepts.

Chapter 7

Consumption and Consumerism

Alternative Perspectives

Sociology has always kept open borders. It has never insisted that practitioners be formally trained and qualified in the subject, and has always been receptive to ideas from neighbouring disciplines – phenomenology in philosophy for example and, in more recent times, geographers' arguments about the importance of 'place' (see Urry, 1995a, 1995b). Sociology's sub-disciplines, through which ideas from neighbouring subjects often enter, are notoriously unstable. Their birth rate is high. The sociology of leisure was an emerging sub-discipline in the 1950s and 1960s. Early on many of its ideas were drawn from the sociology of work. Subsequently, as we have seen, it incorporated arguments from gender and life course studies. These arguments challenged some reigning ideas in the sociology of leisure but it was possible to host the debates within the sub-discipline's established terms of reference, which were to investigate leisure time and spending, their distribution, and how the opportunities and constraints associated with other social roles – occupations, familial, gender and age related – influenced uses of leisure. Debates about the respective merits of contextual (social structural) and internal (social psychological) definitions of leisure were sometimes fierce, but not paradigm shattering.

Currently the sociology of leisure, around which the broader and looser field of leisure studies has developed, faces potentially terminal challenges. The growth of leisure has attracted more widespread scholarly interest. Some of this has increased the number of researchers, writers and students in the study of leisure as already constituted, but the territory has also been claimed by alternative, competitive perspectives. These are considered in this and the following chapter. The now conventional sociology of leisure has three options: to surrender its field, repel the invaders, or assimilate their arguments. It should come as no surprise that this book has not been written to surrender the subject. Its posture towards the invaders is assimilationist, on the basic terms laid down by the now conventional study of leisure as reviewed in previous chapters. It will be argued that the rivals' perspectives are simply inadequate when faced with the full panorama of leisure but that some of their arguments are valid and must be drawn into, thereby updating, enlivening and enriching the field of enquiry.

The new perspectives have various origins.

Marxism

Marxists have always insisted on the basic role of economic processes and relationships but, even so, they have often had much to say about leisure. Marx himself did not write a theory of leisure as such but, as Rojek (1984) has pointed out, it is not difficult to extract a theory of leisure from Marx's works.

In the first half of the 20th century Marxists became embroiled in a series of heated debates about leisure. One set of arguments was about whether bourgeois arts and competitive sport could have any place in a socialist society. By the 1930s there was a clear party line on these matters: the socialist countries would do these things even better than the bourgeois societies in which the activities originated. The authorities in the young Soviet Union displayed intense interest in their people's ways of life. Time budgets were pioneered by Soviet researchers who hoped to chart the breakdown of older divisions such as between town and country, and intelligentsia and workers, and the development of a socialist way of life which would be superior to both the folk cultures of the past and the high culture and popular cultures of capitalist countries (see Roberts, 1990).

Since the 1930s Western Marxists have developed elegant critiques of leisure under capitalism and, equally, in more recent times, of the conventional sociology of leisure that these societies have bred. Marxists have argued persistently that the 'relative freedom' that people possess in leisure is in fact extremely limited; that much leisure time occurs in fragments when it is not possible to do much except recover from work; and that most people have little money left for discretionary spending once they have made all the purchases necessary to participate in their societies' basic ways of life (see Clarke and Critcher, 1985). Marxists have also drawn attention to the legal and moral prohibitions that have been built around people's so-called free time, and how the capitalist state has promoted non-subversive forms of recreation. Here much attention has been paid to the state's enthusiasm to promote sports which have taught young males the virtues of teamwork, discipline and the need to submit to leadership (see Hargreaves, 1982; Hargreaves, 1986).

Box 7.1. The Frankfurt School.

From the 1930s onwards the Frankfurt School became renowned for its critical analyses of how the capitalist culture industries (the press, the cinema, radio, and subsequently television) numbed people's political sensibilities while applauding the Western way of life.

The leading members of this school (see Adorno and Horkheimer, 1977) developed what many still regard as the most systematic, searing and challenging analysis of the culture industries that has ever been produced. It is the original 'class domination' theory in the sociology of leisure.

Its core argument is that during 20th century the culture industries have become pivotal in capitalist domination by creating, then satisfying, false needs. Adorno and Horkheimer argued that the culture industries produced standardized commodities with only superficial variations (films and popular songs for example) which could be sold for profit. The products were designed to command mass attention. Pleasure could be derived effortlessly. People thereby became passive, obedient and uncritical, corrupted by immersion in pleasures that were superficial and ultimately false.

All the members of the original Frankfurt School (see Box 7.1) deplored what they regarded as the overwhelmingly oppressive influence of the culture industries. But some, including Walter Benjamin, believed that leisure time created at least the possibility of workers reflecting on their situations, individually and collectively, and becoming aware of, and resistant to, the restrictive character of capitalism (see Rojek, 1997).

Subsequently Marxists have often dwelt on these contradictory features of leisure. On the one hand, people's minds and emotions are said to be vulnerable to control by capitalist industries. Simultaneously, Marxists have been interested in the working class's struggles to win its own time and space and to develop a culture based on its own organizations – social clubs, brass bands and suchlike. These have sometimes been linked to trade unions and political movements. David Rowe (1995) has argued that all popular cultures, created in space wherein people can pursue their own pleasures, contain elements of 'resistance' to the dominant social order. However, the success of the capitalist industries might be judged by the limited headway of allegedly more 'authentic' working class leisure.

Within Marxism and beyond, the Frankfurt School's arguments have been incorporated within the broader critique of consumer culture that forms the theme of this chapter. Since the 1950s Marxists have had to attend to Western consumer cultures. It became obvious very early in the Cold War era that the leading capitalist societies were giving their working classes higher living standards than the socialist countries. Moreover, the Western working classes were evidently more interested in benefiting from the consumption opportunities offered by capitalism than changing the system. Marxists have been obliged to address these developments. Capitalist societies have not developed in the way that Marxism originally predicted. This has required a Marxist explanation, or the abandonment of the theory, and convincing explanations have needed to treat consumption seriously (see Tomlinson, 1990).

History of consumerism

Social and economic historians have accomplished much detailed work on the development of levels and patterns of spending and consumption during the 20th century. Historians' basic skills are in digging up the past. In historians' work, theory is normally subordinated to evidence if, indeed, any theoretical interpretation of the raw data is considered necessary or permissible. However, some historians have engaged in the debates within both Marxism and the conventional sociology of leisure (see Davies, 1992; Cross, 1993), and the historians' evidence has been available to the other parties.

Cultural studies

This subject did not exist in the 1950s but it has now become a major social science, a spectacular growth area in terms of student interest in the 1980s and 1990s, and has exerted a profound influence in many other disciplines, especially sociology. In Britain, cultural studies began when techniques of literary analysis and criticism began to be applied to the output of new culture industries – the press, radio and television – and to the ways of life of their audiences (see Hoggart, 1957; Williams, 1963). It was cultural studies that first emphasized the cultural richness of working class ways of life. Ordinary people were shown to possess a different culture rather than just less than the middle classes. Henceforth sociological studies of the working class paid as much attention to their values, beliefs, feelings, tastes and forms of sociability as to their standards of living, education and housing. In the 1960s the methods of cultural studies helped to inspire a 'new criminology' that sought to 'appreciate' the delinquent and, subsequently, appreciative (from the inside) studies of working class youth cultures (see Hall and Jefferson, 1976; Mungham and Pearson, 1976).

Cultural studies now exists as a separate discipline, but also within sociology, and within media or communication studies where its speciality is the output of the culture industries. In this work the discipline has developed methods of discourse analysis and ways of deconstructing 'texts' which can also be applied, though with greater difficulty, to the ways of life of the audiences. The study of popular culture is particularly concerned with the interplay between everyday ways of life and the cultural industries which help to sustain them. Since the 1960s cultural studies has been greatly influenced by Marxism which, in turn, has been able to draw insights and arguments from cultural studies.

The sociology of consumption

This is among the newest and, in the 1990s, among the fastest growing sub-disciplines within sociology. This sociological specialism has been able to draw upon both the work of historians and cultural studies' methods of analysis. Otherwise its data are largely the same as in the sociology of leisure: participation rates and spending patterns.

Sociological interest in consumption has been boosted by its relevance to longer-standing central issues in the discipline, namely, class formation and divisions. It has been argued that class formation need not occur primarily in workplaces and occupational associations but may be accomplished through the development of distinctive consumption patterns (see Butler and Savage, 1995). In the 1980s in Britain there was much interest in the alleged formation of a new working class distinguished most clearly by its 'consumption' of housing (owner occupation). More recently, however, attention has settled on the formation of new middle classes which are said to be recognizable, to their members and to others, by, for example, their gentrified inner-city dwellings or Range Rovers and green wellies. Attempts to identify such new classes have led to detailed examinations of food purchasing, the consumption of high culture, wine and so on (see Savage *et al.*, 1992; Warde, 1995).

Postmodernism

This term has appeared in earlier chapters, but at this point it must be given more systematic attention. The topic may seem designed to confuse innocents who will probably not realize that the term has no agreed meaning. Some leading contributors to the debates concede that postmodernism is impossible to define (see Featherstone, 1988). Sometimes the term refers to a set of social and economic trends which are said to be leading to a postmodern condition. Others use the term to characterize modes of social thought whose appearance may or may not be regarded as related to surrounding social and economic changes. There is no unanimity about the character of postmodern thought, economic or social conditions. The sole point of agreement among the postmodernists or postmodernizers is that a transformation is underway or imminent which can be compared in its significance with the earlier birth of modern societies. The study of leisure is inevitably embroiled in these debates because the keywords in postmodernism's vocabulary include leisure, consumption and culture (see Chapter 6, p. 157). Marxists, other analysts of contemporary economic and social trends (especially the rise of consumerism), exponents of cultural studies and the new sociology of consumption, are often parties, though not necessarily all on the same side, to debates about whether we are becoming postmodern.

The significance of consumer culture

This chapter deals with a set of arguments about the rise and current significance of consumer culture. Much of the fine detail is drawn from the work of social and economic historians, cultural studies supplies some of the methods, Marxism (especially the Frankfurt School's analysis) lends a critical edge, and the historical importance of the arguments can be highlighted by adopting the postmodernist vocabulary. Chapter 8 deals with another set of arguments which draw primarily from the sociology of consumption, cultural studies and debates on postmodernism about how leisure can nowadays be used to construct social positions and identities.

The arguments in this and the following chapter offer alternative visions of a postmodern condition, and are not just of major interest to, but in each case represent challenges to, the sociology of leisure as it has developed to date, wherein individuals have been seen as possessing time and money that they can use on goods and services offered by commerce, the public sector, voluntary associations, or self-provisioning. The critique of contemporary consumerism around which this chapter is built alleges that conventional leisure research may have been appropriate earlier in the 20th century. Subsequently, it is argued, leisure has become more thoroughly commercial and pervaded by consumer values that are propagated by, and ultimately serve the interests of the commercial providers. If fully accepted, this critique of consumerism could not simply be absorbed within, but would replace or require a radical reconstruction of, the entire study of leisure as developed up to now. We would need to retrace all our steps, reassess the growth of leisure, work–leisure relationships and everything else.

Consumption

Production and consumption

Debates about consumerism and consumer culture are relatively new but consumption itself is an age-old process. It is as old as work. The two are hand in glove. By working on nature, or on partly produced items, people produce goods and services for later consumption and they have done this in all societies whose members have not simply lived off nature. Consumption is the complementary process. Work increases the value of whatever is worked on. This is why people work. In consumption, value is depleted and eventually used up. People can only consume if they, or others, also work, and an object of working is always to be able, later on, to consume the product of one's own or someone else's labour, or to enable another person or persons to do so. Alternatively, people may now consume prior to working. Credit permits this. The consumers thereby become indebted. But someone must have already worked for there to be anything to consume.

In some types of consumption the value of a product is depleted quickly, for example when a meal is eaten or a beverage drunk. Other goods' value is depleted more slowly. We expect our new cars to last for many years and to have retained some value when they are traded in. The same applies to houses. In fact during some historical periods people have been able to regard their dwellings as investments. Some products can be 'consumed' without diminishing their future availability. A film can be viewed without depriving others of the opportunity. Indeed, the same person can watch a film over and over again. Some works, usually described as 'art', are able to command endless audiences. With most such products, however, individuals eventually tire of repeats, everyone who wishes to watch does so, and the value of the products can then be considered exhausted. By then, of course, the producer or whoever has the property rights is likely to have recouped many times the value of the labour invested in the production. In some cases, like classical works of art, the returns may continue indefinitely. However, these are not exceptions to the rule so much as extreme cases. Normally people work only in the expectation that the value of whatever they produce, whether they are making something for their own use or for wages, will be greater than the value of the effort that they expend.

The separation of production and consumption

Consumption is an age-old process but there are three features which make modern consumption more prominent and important than in earlier times. First, production and consumption are usually, though not always, entirely separate. People still produce things that they intend to use themselves in their homes and may consume them almost instantly, as when meals are cooked and eaten. Nevertheless, what most people, at any rate those in employment, regard as their main work is done for money. Life becomes divided into producing and consuming. People go to work to earn the money to spend in their non-working time. Production and consumption are separated temporally, and also in that people do not expect to consume what they themselves have produced, or helped to produce. They use their earnings to claim shares in the products of other people's labour. Workers earn the money to spend and, in a sense, the time when they can spend it. No one defines the whole of non-working time as leisure. Nor do people spend all their non-working time consuming except in the sense that they can be said to consume their free time. Nevertheless, in modern societies, for most people, there are times and places where they work and other times when they are able to consume.

Consumer choice

Second, in market economies where people work for money they are presented with a huge array of choices and constant invitations to spend and consume. Producers compete to sell their goods and services. Beyond the core value of their products, merchandisers attempt to increase their attractiveness through design, advertising and other marketing strategies. It is these processes that generate a consumer culture; all the images, sounds, smells and tactile sensations which become loaded with meanings and which surround products and the processes of consuming them.

Consumption may seem an odd umbrella term to describe what we are doing with all our purchases and the uses that we make of them. After all, we do not consume furniture and Mediterranean resorts in exactly the same way as pints of beer. But from the point of view of the merchandiser it is all basically the same. Holidays are packaged, furniture is manufactured, then these commodities disappear from the market. The products have left the hands of the producers or their agents and have been appropriated by consumers.

Consumer culture as mass culture

Third, when most people have sufficient spending power to make consumer choices, consumer culture can become mass culture. The birth of mass consumerism is then possible, meaning that some of most people's hopes and aspirations become centred around consumption. Who people feel they are, their social identities and self-concepts, may depend at least in part on what they are able and choose to consume. Individuals' conceptions of the kinds of people that they would like to be, and try to become, may be constructed in terms of consumption.

The extent to which all this has in fact happened is, of course, among the key issues for debate. The following passages consider both sides of the relevant arguments. However, there can be no question that in modern societies consumption has grown both quantitatively, in terms of the sheer amounts that people are able to spend and consume, and also in its cultural significance.

The Construction of Consumer Culture

Beginnings

Consumerism is a recently coined, post-Second World War term, but the phenomenon has attracted attention for much longer (see Box 7.2).

Box 7.2. The affluent worker in the 19th century.

When John Goldthorpe and his colleagues (1969) conducted their study of affluent manual workers in Luton in the 1960s amidst a debate on the alleged embourgeoisement of the working class, they discovered that there had been a similar debate almost a hundred years earlier.

As Chapter 2 explained, in the closing decades of the 19th century workers were earning higher wages and enjoying higher standards of living. They were becoming better housed, taking holidays away from home at expanding seaside resorts such as Blackpool, and there were new forms of entertainment – spectator sports, music halls and theatres – for those able to afford the admission prices.

People's leisure opportunities expanded in the early years of the 20th century, and especially between the World Wars when the radio and the cinema developed rapidly as forms of mass entertainment (see Chapter 2). Driving and flying also became new leisure activities, for the relatively well to do, but cycling was the mass pastime. Simultaneously, new forms of public transport enabled more people to take day trips and longer excursions to coastal resorts and the countryside. Rambling thus became another popular recreation. The holiday resorts continued to expand and professional football clubs built new, larger stadiums. Among young people, modern dancing became popular, but it was developments in the manufacture and purchase of consumer durables that may have had the greater significance.

New forms of marketing

There was much excitement when department stores began to open from the mid-19th century onwards. These attracted the same kinds of attention as supermarkets in the 1960s, and shopping malls and retail parks subsequently. It was noted when department stores first opened that these were often people's main targets during days out. There was much discussion and concern about shoppers, especially women, becoming bewildered, dazzled, seduced and tempted to spend beyond their means by the array of goods all within their physical reach but usually beyond their immediate means. In the 19th century the department stores catered almost exclusively for middle class shoppers, but subsequently aimed at a wider market.

During the 1920s and 1930s, mass markets were developed for a new range of consumer products which were displayed prominently in the department stores and elsewhere. Some of the new products, or at least their mass marketing, became possible as more and more homes were connected to mains electricity. Irons, vacuum cleaners and lighting fixtures fell into this category. Gas fired boilers and cookers, and radios and gramophones were other goods which sold in mass quantities. At that time refrigerators remained luxury items and washing machines even more so.

In retrospect sociologists have traced the beginnings of home-centred lifestyles to the inter-war years. This was the period when manufacturers began to use advertising to create or structure mass markets for their particular products (Bowden, 1994). In America, Henry Ford, then numerous competitors, began to mass produce motor cars with the aim of bringing their price within the means of the average family. This was achieved by mechanization and mass production which meant, for example, that customers could have any colour 'so long as it is black'. In Britain it was not until the 1960s that private motoring became a mass phenomenon.

Enduring divisions

The inter-war years are, of course, best remembered for the recession, unemployment and hard times that blighted the economic and social landscapes, but, for people who remained in work, rates of pay and standards of living rose. The new types of consumption developed most strongly and spread most widely in southern England rather than in regions where levels of unemployment were much higher. And even in the more prosperous parts of Britain the majority of people were not part of the consumer boom. Until the Second World War around one-third of Britain's manual working class lived beneath or barely above the poverty level, meaning that their incomes were sufficient to purchase little more than necessities. The typical working class family could expect to sink into poverty at vulnerable points in the life cycle, particularly when one wage earner was supporting a spouse and children, and later on in old age. Lay-offs and unemployment were likely to create financial crises at other times. For most working class households life was a constant struggle to maintain decent standards. Respectable working class families might have placed themselves above the 'roughs' but they did not share middle class lifestyles.

Many working class families were rehoused from inner-city slums to council housing estates while middle class families were purchasing the detached and semi-detached properties that were being built throughout the length and breadth of Britain. At that time middle class children were educated at different schools – secondary schools – while most working class children received elementary education only. The middle classes worked in and dressed for office environments, and had their own leisure tastes and activities. When the working and middle classes went to the same places for entertainment – to football grounds and music halls for instance – they usually stood or sat in different areas (see Mason, 1994).

Within all social classes men were the normal breadwinners and there were huge differences between males' and females' ways of life. There were also pronounced age differences. Successive cohorts of males and females followed the life cycle sequences and patterns characteristic of their social classes. Consumption rose in all these groups, but, prior to the Second World War, not sufficiently or widely enough to breakdown the characteristic group patterns (see Davies, 1992).

The breakthrough

Social historians trace consumer culture's decisive breakthrough to the period between the 1930s and 1950s. The Second World War, of course, bisected this period, but was more important as an accelerator than a catalyst. In retrospect it is possible to identify a set of interrelated key developments.

First, a new economics associated with John Maynard Keynes explained how governments could spend their countries out of recessions and mass unemployment. The USA's 'new deal' programmes in the 1930s applied this prescription. European governments followed suit as they began to re-arm. Needless to say, during the War itself all labour was needed. By the end of the War there was inter-governmental agreement that in the new world order mass unemployment should never be allowed to return. All Western governments adopted Keynesian economic policies and the countries experienced several decades of full employment and economic growth.

Second, industrialists and their spokespersons endorsed this type of state intervention. They remained hostile to 'socialism' – meaning nationalization and high taxation – but not to 'sensible' macro-economic management. Expanding economies which kept workers in jobs with progressively higher earnings meant larger and wider markets from which businesses could profit.

Third, once full employment was seen to be a realizable goal, trade unions softened their demand for shorter hours of work. They demanded full employment policies of governments and, from employers, progressively higher earnings (see Hunnicutt, 1988).

Fourth, the population's ability to spend and consume grew, and people's appetites for still higher levels of consumption were stimulated. Married women began entering the labour force in increasing numbers from the 1940s onwards and, unlike during the two world wars, this time they were in the labour market and staying. At first many explained that their earnings were for extras and that their husbands remained the real breadwinners. Everyone has subsequently learnt how one decade's luxuries become normal expectations in the next. Today life would be intolerable for many households without central heating, televisions, vacuum cleaners and telephones.

Consumerism as the dominant culture

Consumerism takes root as spending becomes increasingly central in people's life goals. Gary Cross (1993) has argued that this had happened by the 1950s throughout the Western world and that consumer culture has continued to reign since then. Likewise Celia Lury (1996) presents consumer culture as the material culture that has been dominant in Euro-America throughout the second half of the 20th century. People have been willing to work hard and long, and in recent years, in some cases, even harder and longer than previously, in order to satisfy their consumer aspirations. Simultaneously, more and more goods and services have become commodities which are produced for and purchased by more and more people.

In consumer societies governments are judged by their ability to maintain rising standards of living. Trade unionists judge their leaders by much the same yardstick. This has remained so despite the return of mass unemployment since the 1970s. Of course, politicians of all parties, and certainly the trade unions, want and demand a return to full employment, but in practice all these leaders are more responsive to the demands of the mainly employed voters and fee paying trade union members who continue to judge their leaders according to the trends in their pay packets. Even though opinion polls may suggest that voters regard unemployment as a leading issue, if not the leading issue, the 'feel good factor' continues to depend more on trends in the self-perceived prosperity of the majority who are in jobs.

Since the 1970s their real earnings and living standards have continued to rise. They have been responsible for the rise in spending on video-recorders, cable and satellite television, long-haul major holidays, second breaks, and so on (see Martin and Mason, 1986, 1990, 1992). The interests of businesses, the earn and spend aspirations of the workforce, and the electoral pressures which all political parties that aspire to office must respect, have formed a self-sealing circle. However, Gary Cross (1993) has warned that we should not expect the consumer culture to reign indefinitely. It has been dominant in the West for only half a century and is unlikely to be so until the end of history. Whether the consumer culture is currently disintegrating, or threatened, is an issue to which we shall return shortly.

For the present it should be noted that some sections of the population have remained outside the consumer society. In Britain about one-third of the population take no holidays, not even a single night away from home, year in and year out. A similar proportion are without access to private motor transport. But today when people talk about a two-thirds, one third society it is the two-thirds who are the alleged 'haves'. This is unlike the 1930s and before when poverty was a general working class problem and the working class amounted to around two-thirds of the population. Today's 'have nots' are not necessarily suffering the same kinds of hardship as those in poverty before the Second World War. However, the poor today are excluded from most of the consumption that is taken for granted by most members of their society. This is a different kind of deprivation which, in its own ways, may be just as painful as the poverty of earlier decades.

Shopping and Leisure

The monetarization of leisure

In a consumer society one would expect shopping to be a core, if not the core leisure activity, especially if shopping is defined broadly to include all kinds of looking at and for things to buy, preparing to purchase, and subsequently not only using the purchases but also reflecting on the value of the deals and considering what to buy on the next occasion. All this can be expected to supply peak leisure experiences for people who are immersed in consumer culture. This should be reflected in the time that people devote to shopping, and its position in their overall leisure schedules. Chapter 3 explained that leisure time has expanded much more slowly than people's spending power in recent decades. Consumer spending almost doubled in Britain in the 25 years up to 1996 while free time hardly increased at all (Martin and Mason, 1998). Therefore, merely purchasing, not to mention using the purchases, accounts for a growing proportion of people's non-working time. Time budgets have recorded the increase in the amount of time spent shopping in Britain; up by around 50% since the 1960s. All this is what is meant by the 'monetarization of leisure'. More and more of our leisure time has become geared to purchasing.

Commodification

More and more goods and services – everything that people want or can be persuaded to want – are packaged as commodities to be acquired by anyone who will pay the price. Holidays are bought as packages. Excitement and exotic experiences can be bought by visiting the appropriate theme parks. Friends of a preferred type become available by joining the right contact club. Conversation is available for the cost of a telephone call. In societies where all this is possible one would expect people's aspirations to become geared to spending. All types of experiences become things to buy. Everything has a price. So what most people want above all else, and what they feel they really need, is money.

Ubiquitous marketing

Merchandisers have become extremely imaginative in marketing their wares in order to meet or preferably structure the public's demands. Much of the country's most creative talent has been recruited into advertising. It is no longer sufficient to have a product stocked on shop shelves. Goods and services need to be designed, packaged and promoted to make people aware of their availability, and what the purchases will do for them. Gas cookers, blue jeans and many additional otherwise mundane objects have been surrounded by all kinds of improbable imagery. Developing and disseminating the imagery has become a high status, well-rewarded activity. Leading actors who once feared that appearing in adverts would damage their reputations are now keen to do voice-overs. The recipe obviously works. Consumers are said to acquire not just the material products but also the associated meanings. Being seen in the right restaurant, night club or holiday resort, and wearing the right brand label on one's shirt or trousers, advertises an individual's membership of 'people like that'.

The sums now spent on consumer market research dwarf the budget for academic leisure studies. Nowadays consumers' motivations and attitudes are explored assiduously. Most of the findings never become public property. They belong to the commercial purchasers. Great expense is incurred in order to discover the brand name, the shape of car and so on that will result in marginal but often crucial increases in sales.

Marketing is ubiquitous. It is impossible to move through cities or the countryside, or to open newspapers or surf the television channels without being reminded of purchasing possibilities (see Wernick, 1991). In a consumer society the most sensitive indicators of trends in leisure, more sensitive than time budgets, are data on spending. This shows, for example, that during the 1980s and 1990s the main growth areas in UK consumer spending were on overseas holidays and home entertainment equipment. Other leisure commodities, including alcohol, did not benefit in the same way from the overall rise in consumption (see Martin and Mason, 1986, 1990). Irrespective of whether our self-concepts are now based primarily on what we buy and where we buy it, our uses of leisure are increasingly based on or at least geared to purchases.

Chore or leisure?

It is difficult for leisure researchers to decide how to treat shopping. The customary practice in time budgets has been to classify shopping as one of the chores that simply has to be done, but this now seems to be at loggerheads with where, when and how people shop. Shopping might be used as a leading example of a postmodern condition in which modern divisions and categories are blurred. Work spills into what used to be leisure time as more people take work home, or go to work in the evenings and at weekends, while people find leisure experiences in their paid jobs and in what were once unpaid chores. Times, places and activities become impossible to classify unambiguously as work, unpaid chores or leisure. The only sensible question may be how people experience the times, situations and activities (see Box 7.3).

Box 7.3. Attitudes to shopping.

People's attitudes to shopping vary enormously, from love to hate.

People often explain how their feelings depend on the type of shopping. Whereas the weekly supermarket trip for groceries may be experienced as a chore, people often feel differently about shopping for a new outfit.

Men and women today may spend similar amounts of time shopping, but attitudes are still gender divided. The people who love shopping are mostly females while those who hate it are mostly men. Hence the case for treating shopping as the female mode of consumption *par excellence* (Campbell, 1997).

Shopping was once a woman's chore. The fact that men now spend almost as much time shopping as women perhaps suggests a change in the activity's meaning. When people have days off work shopping is often their core activity. The shopping is not done as quickly as possible and put aside to leave as much time as possible for other uses of leisure. Saturday is the week's main shopping day. When shopping itself is treated as a leisure activity it becomes far and away the most common form of out-of-home leisure. The introduction of Sunday trading in Britain in the 1990s immediately transformed the Sabbath into another major shopping day. Sunday shopping quickly became so popular that stores and car parks became congested and traffic jams extended for miles despite the absence of much of the normal weekday work traffic.

The shopping environment

The retail trade knows full well that simply pricing goods down and making them available will not always maximize sales. Shoppers are attracted to complexes with numerous stores through which they can browse, and where there are places to relax and refresh. Some complexes contain entertainment and play areas for children (and adults). Individuals, families and organized groups travel for miles to major shopping centres. In the USA the cities' shopping complexes are often central attractions in the advertised weekend breaks for which people travel in a variety of ways including by air. Of course, shopping has always been a part of the traditional holiday. The main resorts' attractions have always included their shopping facilities. Nowadays, it seems, the shops are often the main attraction. Hence the spate of recent enquiries into the types of facilities that tempt people to shop as a leisure activity (see Jansen-Verbeke, 1987; Roberts, 1987; Shields, 1992).

Purchasing has become time demanding because prior to spending their money many shoppers like to view all the alternatives. Window shopping is a normal prelude or accompaniment to actual purchases. Major and minor purchases may be surrounded by extensive sampling often in the company of friends or family members who can offer authoritative advice. The occasions then supply much material for subsequent small talk. Friends and families discuss who has bought what, together with their future spending hopes and plans.

Major shopping developments are now consciously planned and designed to appeal to the leisure shopper. Pedestrianization is known to enhance the quality of the experience judged by the effects on trade. Maximizing pulling power depends on having the right mixture of large and small stores offering different kinds of wares. More conventional leisure facilities – cinemas, gardens, pubs, restaurants and art galleries – can add to the appeal. Conventional tourist attractions are now used routinely to tempt shoppers. Some resorts have natural environmental features that would be blemished by over-development. They need to strike the right balance but they can be sure that today's visitors will have cash to spend not just on accommodation and food but on virtually everything else that the stores can offer.

Sporting spectacles are now sought by cities throughout the world. The value of the television rights is important but so are the visitors who will pay admission prices and pay for many other things as well. However, the arts and heritage are unrivalled for attracting tourists. They can attract visitors throughout the year, year in and year out, and the same visitors will sometimes visit again and again. Moreover, the types of tourists attracted by the arts and heritage tend to be people with money. The core attractions can be made available free of charge. The cities profit from visitors' spending while in the vicinities. The modern tourist can be relied on to shop. In a consumer society all major leisure occasions – holidays, Christmas and family celebrations – become orgies of spending and consumption.

New forms of shopping

It is unnecessary to go out in order to shop. Daily and weekly reading matter is packed with adverts. So are most television channels. In fact there are dedicated shopping channels on cable and satellite TV. The adverts on other channels, and in the press, do not seem to repel readers and viewers. They are probably an added attraction. However, it is not only the explicit adverts that persuade people to spend. News features and photographs, and television programmes, cannot help but display products that people decide they would like to possess. And the purchasable items are often shown alongside attractive role models. Ellis Cashmore (1994) has argued that a main effect of television has been to make people spend more. It may be unclear whether television sex and violence spread these phenomena into real everyday life but there seems to be no doubt that exposure to television makes people spend. If they do not have cash in hand or in the bank there is now a massive credit industry to take the waiting out of wanting. Like many features of the consumer society, hire purchase was introduced in the inter-war years and has since become a normal way of acquiring not just durables but also items such as clothes and holidays.

For a long time it has been possible to shop by mail or telephone. It has been predicted that before long we will be able to call up demonstrations on our computer monitors before sending orders 'down the line'. However, it seems unlikely that people will be lured from the attractions of live shopping. The new alternatives are most likely to be added to, without replacing, the traditional activity.

The promotional culture

Even people who are unable to purchase cannot escape the consumer culture. According to Wernick (1991) we all now live amidst a promotional culture in which advertisers' clever techniques have to be copied by everyone – politicians and even academic authors – who wants to deliver a message. No serious political party would now face an election without its 'spin doctors'.

There is no doubt that consumption and consumer culture have grown in scale and importance. All this is generally agreed. The controversial issues are, first, whether consumerism has become so large and powerful a force that it excludes, dominates or marginalizes other uses of leisure and, second, whether consumer culture results in the manipulation of consumers by commercial interests and therefore by a capitalist class, as the Frankfurt analysis suggests, or whether, as orthodox economic theory claims, the customers are ultimately sovereign.

The Public and Voluntary Sectors in a Consumer Society

Commercial leisure, which provides goods and services for profit, has been the central force in the spread of consumer culture, but other kinds of leisure provision have inevitably been affected. The public and voluntary sectors now operate in a different context compared with when most of the services and many of the organizations had their origins.

Traditional public provisions

Chapter 1 explained that the public sector will provide whatever politicians want, or are willing to endorse, subject only to normal political constraints. Whatever their private motives, public providers have usually claimed to be catering for people's needs. In practice, however, it has usually been the providers' conceptions of other people's needs that has been decisive. Traditional public provisions have been basically paternal.

The providers have usually claimed to be making sport, the arts, the countryside and so on accessible to everyone, thereby widening their opportunities. They have often stressed their responsibilities towards otherwise disadvantaged groups: children and young people, the poor and, in more recent times, women and ethnic minorities. In practice, however, the provisions have never been used by everyone. And public provisions have rarely made much impression on social inequalities in participation. In sport, for instance, social class differences in participation remain virtually as wide as ever, and despite the sustained promotion of classical music by the BBC since the 1920s, the audience has not grown significantly, nor become more representative in its social composition. Sustaining traditional public leisure provisions always depended most of all on the providers' ability to satisfy key constituencies – the vocal and politically influential – rather than the public in general. For a long time most citizens were acquiescent but the spread of consumer culture has changed this.

Public leisure services in a consumer context

There are several ways in which consumerism and related developments have undermined paternal public provisions. First, in the past, except when they went away on holiday, most people had little choice but to use local leisure facilities. Today the public is more mobile, thanks mainly to the spread of private motor transport. People have a wider choice of destinations when they go out for an evening or at the weekend. Local authority leisure facilities now need to compete with public and private providers in neighbouring areas if they are even to hold on to their local publics. And this usually means offering what people manifestly want rather than what they are believed to need.

Second, in the early days of the modern public leisure services, there was simply no way in which most people could afford to play sport, visit museums and art galleries, and to use parks, except by using free (at the point of use) or heavily subsidized public services. Nowadays, in contrast, more people can afford to pay for, and are offered, commercial options – theme parks, tennis centres, and exhibitions of various types, for example. The effect, once again, has been to loosen the public sector's hold over its traditional markets.

Third, as people have become more accustomed to paying for leisure, and as more and more opportunities have been made available by the private sector, citizens have inevitably begun to ask why they should pay for public provisions that they do not use. As a result public providers are finding it more and more difficult to increase their budgets from tax revenues. Politicians start to demand efficiency gains and suchlike; they also start to demand clear objectives and measurable outcomes. Even to maintain their resource levels, public providers are likely to discover that they need to seek revenue in alternative ways, by seeking sponsors and charging users for example. In other words, they experience pressures to become more commercial.

These are the developments that underlie recent changes in the ways in which public leisure services are administered. In 1992 in Britain local authorities became obliged to introduce competitive tendering for the management of most of their leisure services. So it has been possible for commercial and voluntary organizations to take over the management of these facilities. Up to now, in most cases, the contracts have been awarded to in house bidders, but even so, under the new regime the providers have been under pressure to cut costs to a minimum and to maximize revenue from outside the public purse. This has usually meant catering for people's manifest wants; in other words, becoming more commercial. Since 1997 compulsory competitive tendering has been replaced by a 'best value' requirement, but the effects are basically the same. Whether these are applauded or deplored depends largely on commentators' assessments of the respective merits of traditional public and commercial leisure provisions. Some believe that the recent changes will leave the public better served (Coalter,1998). Others deplore the further spread of consumer-based capitalism (Ravenscroft, 1998).

Voluntary associations

These have been able to continue in their customary ways, provided they have been able to attract sufficient members with sufficient enthusiasm. In practice, however, with present-day consumers having access to a wider range of easier attractions, many clubs and societies have experienced a loss of volunteers. This has affected youth organizations, sports clubs and similar bodies.

Simultaneously, some voluntary organizations have had the option of becoming more commercial. This has meant some combination of introducing commercial-type marketing, raising user or participant charges, seeking sponsorship, and appointing paid staff, which is how segments of sport and the arts have been reshaped into commodities.

The Critique of Consumerism

Peter Taylor (1992) has complained about an anti-commercial bias in leisure research. He has a point. His charges are that commercial leisure has been either ignored despite its obvious importance or disparaged without sufficient evidence to justify such treatment. It is certainly the case that there have been hundreds of academic studies of public sector policies, provisions and participants, but very few equivalent enquiries into commercial leisure. It is also the case that most academic writing has been critical of the introduction of commercial pressures into Britain's local authority leisure services. One reason for the neglect of commercial provision, of course, is that commercial providers are more likely to commission their own market research than to call on academics. The latter have found the public sector far more willing to consider their research findings and conclusions carefully. However, very few leisure researchers have displayed a desire to undertake sympathetic studies of the situations and problems of commercial operators.

The corruption of pure leisure?

The academic literature is littered with criticisms of consumer culture. This is often regarded as defiling or deforming true leisure. Writers construct images of past times, sometimes invoking the ideas of Aristotle and Plato, when a superior kind of leisure is supposed to have existed. Classical contemplative leisure has been compared favourably with the hectic pace of modern consumption (see Dare *et al.*, 1987). Commercialism has been accused of filling people's free time with mere amusements and diversions (Hemingway, 1988), creating endless insatiable desires that leave people unfulfilled and restless (Goodale and Godbey, 1988; Wearing and Wearing, 1992), and turning people into passive receptacles of entertainment to which no critical response is possible (de Grazia, 1962).

An inescapable problem with commercial leisure, according to these arguments, is that it requires a division between the provider and the consumer. A consequence is said to be that the consumer's experience is inevitably more limited than when people create their own leisure either individually or in groups. The contrast here is between, for example, the family group which might once have played music on a piano and sung along, and the modern audience for music which is far more likely to listen to records, tapes or discs. It can be argued that there are special satisfactions in self-provisioning.

> We believe that involvement in such groups (voluntary organizations) offers people something probably unique in our society: the chance to come together with others to create or participate for collective benefit and enjoyment rather than for sale to an impersonal audience or producer.
> (Bishop and Hoggett, 1986).

Harper (1997) has argued that leisure is only satisfying when it involves some work-like features: that true happiness really does have to be earned. He criticizes much contemporary provision for making leisure too easy and offering only hollow satisfactions. Jeremy Seabrook (1988) noted, in his interviews exploring the experiences of people from several parts of Britain, that their most meaningful activities did not usually include those purchased in the market.

People who make their own leisure can be creative and experience growth in their skills and capabilities. Equally important, they remain in full control of the character and pace of their activities. Serious leisure (see Chapter 6) appears always to be based on the arts, crafts, hobbies or volunteering rather than purchases of commodities (Stebbins, 1992; Parker *et al.*, 1993).

Undermining alternatives

Commercialism is said, by its critics, to undermine or suppress other kinds of leisure. These are likely to be marginalized and devalued by the culture that surrounds consumption. Spending confers status. Moreover, provided people have money it is very easy to spend to be entertained and amused. People are offered endless opportunities. Purchasing a holiday package is more straightforward than constructing one's own itinerary. Operating a cassette player is easier than learning to play a guitar or to sing. People who are reared in consumer societies may lack the basic skills, or even an awareness that there are alternative ways of creating enjoyable leisure. They may feel unable to have holidays if they cannot afford a package. Withdrawal of television or telephone communication may be experienced as a crisis. Hence the argument that wherever commercial forces are unleashed, other kinds of leisure become casualties.

Individualization

A related criticism is that consumerism tends to break-up communities. The contrast here is between the bonds that develop among people who make their own leisure and between people who purchase similar goods and services. Victoria de Grazia (1992) has made this contrast:

> the old ideal, for better or worse, was the politically mobilized citizen occupied in uplifting activities, whereas the contemporary ideal has become the individual consumer, self-interestedly and occasionally occupied with others, choosing his or her leisure activities at will, almost invariably by commercial means and influenced by a commercial idiom. . . The new associationalism differs in several respects from the conception and practice of organized leisure in the past. It caters to a wide spectrum of clienteles, defined less by class than by market segments formed around income, gender, age, ethnicity and taste.

Post-communism

In recent times there has been a unique opportunity to study the impact of commercialism on people's leisure as a result of the collapse of communism in the former Soviet Union and East-Central Europe. The pace of change has been rapid; the societies became open to Western observers in the early 1990s when older ways of life were still available for study as were new emerging lifestyles which involved the purchase of the consumer products that began to flood in from the West. Under communism there had always been chronic shortages. Most goods were in short supply. People were told that they were fortunate to be experiencing the development of a socialist culture which offered superior satisfactions to the consumer role (Hidy, 1982), but this was probably insufficient compensation to people who were spending hours in queues. As soon as market reforms were introduced the queues disappeared. The city stores became stocked with the full range of Western products. It became possible for consumers to obtain whatever they wanted – cars, fashion clothing, video cassettes and so on – provided they could pay. What was being lost?

First, the lifestyles of the former intelligentsia were being undermined. These strata were at risk of disintegration and their members were certainly losing their positions as high status, high profile role models. The lifestyles of the intelligentsia were formerly based partly but crucially on the production and consumption of state subsidized high culture. With the end of communism most of the subsidies were withdrawn. The incomes of most intelligentsia families were insufficient to pay market prices for their cultural tastes. So domestic cultural production declined and was replaced by imported Western culture, mainly the low brow varieties (see Jung, 1994).

The earlier ways of life of other strata were also at risk. Under communism people retained the art of spending time without spending money. They had no alternative because their incomes were low and, in any case, there was little to buy. Young people would spend afternoons and evenings with friends in each other's homes, on the streets or in local parks. At weekends, and sometimes for holidays, they would visit nearby lakes or mountain regions, taking not only tents but often sufficient food to meet their needs. All this would be done while spending little money, if any. Young people in the West would probably find this impossible. It appears that they cannot go out, into town centres or even down their neighbourhood streets, without purchasing confectionery bars and cans. Western youth believe that they need fashionable jeans, shirts and trainers before they can face their peers. They also feel that they need the latest recordings for their home entertainment. Young people in the communist countries have succeeded in appearing much the same as Western youth, but at a much lower cost. They have also been skilled at keeping abreast of the latest sounds by making and exchanging pirated tapes.

These skills, and the ways of life in which they were embedded, were not lost as soon as market reforms began, but they were clearly at risk (see Roberts and Jung, 1995). A minority of families were able to benefit immediately from the new earning opportunities that became available in the expanding private economies of the former communist countries. They were able to consume conspicuously and become the new high profile, high status role models. The advertising that accompanied market reforms extolled the lifestyles in question. And the new capitalist businesses with Western links demanded the suppression of the pirating and copying of their products. By the mid-1990s a visit to McDonald's had become far more attractive to young people in East-Central Europe than an evening in the local park. The kind of communalism that had been sustained when people had to make their own leisure, and which 'insulated' young people from the effects of unemployment in the early 1990s (see Chapter 3), appeared likely to become another casualty of the reforms.

Commercial manipulation

The final part of the critique argues that consumerism has not triumphed in response to the authentic demands of customers but through the power of commercial interests. As explained at the beginning of this chapter, since the 1950s a variant of Marxism has switched its focus away from the relationships of production to the relationships of consumption in explaining how, in present-day circumstances, the non-capitalist classes become locked into the system. These critics of consumerism have noted that consumption is just as vital as production in sustaining a capitalist economy. People need to be induced to sell their labour power. Equally they must be persuaded to spend their earnings. A strike of consumers would be just as crippling to capitalism as a strike of producers. Needless to say, a consumers' strike is far less probable. Consumers are dispersed whereas workers are drawn together in workplaces which create opportunities for them to form conceptions of common grievances and interests, and to organize for collective action. Hence, it is argued, the crucial role of consumption and consumer culture in sustaining capitalist systems.

It may not always be easy to make people want to work. It is usually far simpler to make them want to consume. And once they want to consume they need to work and earn. Once consumer wants have been created capitalist systems can operate with few overt controls. The systems become 'hegemonized'. Capitalism is made to appear a natural response to people's wants. Early capitalism succeeded by controlling labour whereas under late or postmodern capitalism it is the control of consumption that is said to be crucial.

Marxists, needless to say, have remained critical of capitalism. They have insisted that workers get poor value by selling pieces of themselves when they work daily then try to buy them back in the after-hours in the form of fun (Wright-Mills, 1956). They argue that while consumers can choose which particular products to purchase, they cannot determine the range that is made available, which always confines them to the consumer role (Clarke and Critcher, 1985). Furthermore, they point out that the price of making consumer choices is acceptance of the ideology of consumption, which is how people become locked into a basically exploitative and oppressive system (Chambers, 1983). According to Baudrillard (1998), this lock-up is consolidated by the consumer industries which surround the entire population with a profusion of goods, thereby creating an illusion of mass happiness and equality. Anyone who does not share these feelings is made to feel odd rather than normal.

'McDonaldization'

The basic arguments in the critique of consumerism were originally formulated by the Frankfurt School but have been extended from the original culture industries to the wider range of consumer industries and the marketing techniques that have developed during the second half of the 20th century. In addition to the original Marxist version there is now a Weberian variant which highlights the progressive rationalization or 'McDonaldization' of leisure and society in general.

Ritzer (1993, 1998) has argued that Weber was basically correct about modern societies being progressively rationalized but mistaken in believing that this would be principally through the spread of large bureaucracies. He argues that the form of rationalization that is becoming dominant is in fact best illustrated by McDonald's. This business operates on the rational principles of efficiency, calculability, predictability and control, and pushes them to their utmost. Every aspect of production and delivery is measured. All the burgers are a precise size and are cooked for a precise period. They are exactly the same wherever they are sold. The surroundings, the restaurants themselves, are almost as predictable. Ritzer argues that this kind of rational organization is spreading rapidly in education and medical care, as well as in sport and leisure activities in general. He recognizes, following Weber, that rational organization can produce irrational consequences such as boring products which consumers may desert for something more exciting. Nevertheless, on account of their efficiency, and the predictability of what will be obtained at what cost, Ritzer believes that McDonald's have a decisive advantage. Hence his prediction of a progressive McDonaldization of society.

Resisting commercialization

Some critics of consumerism have formulated a reform agenda aimed at rescuing consumption from capitalist market relationships and making it expressive, passionate and liberating (Tomlinson, 1990). There are pessimists who believe that consumerism is irresistable and will roll on remorselessly. Some consider parts of the modern consumer industries to be beyond reform. For example, Mander (1980) has advocated the elimination of television, and insists that this case be taken seriously because, in his view, no reforms could address television's basic flaws. In Mander's view these are to deny individuals immediate experience by virtue of the ease with which second hand experience is made available, elite control, the creation of confusion and submission among viewers who are bombarded with powerful but often contradictory messages and images, and the inability of the medium to convey anything except simple linear messages. However, other writers have made a case for step-by-step reform (see Tomlinson, 1990).

They hope, first, that through exposing their ideological role the influence of the capitalist consumer industries will be weakened. Simultaneously, they advocate making as many of these industries as possible democratically accountable, meaning responsible to their workers or consumers. They also propose strengthening the voluntary associations through which people are able to generate and manage their own leisure opportunities. The intention is to provide more satisfying alternatives to capitalist consumption and to enable people to form relationships with one another and, hopefully, to acquire the skills with which to take the class struggle into other domains such as work and politics.

This in fact was the agenda of the Greater London Council (1986) in the early 1980s when it sought to redirect some of its leisure spending towards tenants', women's, gay and lesbian groups, ethnic minority and other community organizations. The extent to which the Council's leisure spending was shifted away from the usual (mainly middle class) beneficiaries was in fact minimal, but this did not prevent the Council's policies being labelled 'loony left' and featuring prominently among the arguments that led to its abolition (Tomlinson and Walton, 1986).

The Defence of Consumerism

Leisure and consumption

Is radical reform necessary? Is consumerism as hostile to satisfying leisure as its critics suggest? No one denies that consumption has grown and has become an important part of leisure, but claims that consuming and consumer values now dominate people's leisure are controversial; inevitably so because they ignore huge bodies of evidence on how people spend their leisure time. The evidence and arguments of the critics of consumer culture are not wholly wrong but, in my view at any rate, they are best employed alongside or upon the base laid by more conventional studies of leisure. The critique of consumerism is an example of how focusing on just one part or aspect of people's leisure can give a distorted picture of the whole. It is true that nowadays most people purchase their holidays, Christmas celebrations and evenings out, and that when they spend leisure time at home they typically use consumer goods such as televisions, telephones, soft furnishings and so forth. But does this mean that leisure is being reduced to, or dominated by, consumption?

Graham Murdock (1994) has reminded us that a substantial minority is excluded from most parts of the present-day consumer society. In the UK the numbers of unemployed and retired on state benefits have risen in recent decades, and many members of these groups have experienced a decline, not a growth in their consumer spending. The spending of many households is swallowed by essentials. They are adrift from the strata whose members are able to spend the winter months planning their summer holidays, who spend the preceding months spending towards their Christmas celebrations, who scan the colour supplements in search of purchasing suggestions, and who can afford to go lifestyle shopping on even occasional Sundays.

Even among the relatively affluent it is misleading to portray their entire leisure in terms of consumption. It may be the case that in most of our leisure time we are using consumer products, but this is not all that we are doing. We can be simultaneously relaxing, thinking and talking. And people do not simply consume but *use* most of the goods that they buy. Consumption has become an over-worked concept. It may all be straightforward consumption to the suppliers but it is rather different from the purchasers' perspectives. John Kelly (1986, 1991) has pointed out that in all social strata the most popular leisure activities in terms of time accounted for are relatively low cost, and that most people's core activities revolve around social relationships rather than bought goods or services. Kelly has shown that when people prioritize their own uses of leisure they insist that the activities in which they participate with other people are more important than the commodities that they acquire and possess. This was certainly the case among the 17–19 year old Finns in Pontinen's (1996) study (see Chapter 1, page p. 23). They spent a great deal of their leisure time listening to music on the radio and tapes, and watching television. However, their leisure priority, to which everything else was subordinated, was spending time with their friends.

Leisure is broader than consumption conceptually (see Mullett, 1988) and in terms of the reality of people's everyday lives. There has been a growth in consumption and commercial leisure provision, but these trends have *not* squeezed other kinds of leisure. Much leisure time is still self-organized privately by individuals, families and groups of friends. The voluntary sector has not diminished but has actually grown, only parts of it have been commercialized, and the strata with the highest incomes and spending power are the most likely to belong to these organizations. Also, the public sector has played and still plays a major, not a residual role in leisure provision, and although it is now subject to greater commercial pressures than in the past its sustenance still depends primarily on political will, not market forces.

Consumer choice

Portraying consumers as passive victims whose purchases and leisure are governed by the skill of advertisers and other marketers conflicts with what we know about how purchases are actually made and used. The evidence suggests that customers are usually discerning and discriminating rather than gullible, and typically use the things that they buy for their own purposes. People purchase motor cars so that they can then drive themselves to and for leisure, to places and at times of their own choice. They buy garden tools so that they can design and cultivate their plots according to their own tastes. They purchase standard 'boxes' to live in and then paint, decorate and customize the dwellings according to their own aspirations. Are these choices trivial? Are they from a very limited range of options? Has the range been narrowed or has it widened with the growth of the consumer industries? The development of commercial leisure during the 20th century in Western countries has surely been a great liberating force which has done more than anything else to widen people's leisure options.

There is now a considerable literature showing that most media audiences are anything but passive. In addition to audience research that calculates the size and composition of the audiences for television and radio programmes, and the readerships of newspapers and magazines, we now have studies of how people actually view and listen, and what they do with their reading matter. This research shows, for example, that television viewers are not passive receptacles. Few can be described as hooked or addicts. Families discuss and sometimes quarrel over which channel to select. The action and talk on screen may lead to discussion and argument among the viewers (see Morley, 1986). This is in line with the findings from early research into the impact of the printed media, about a two-step flow of communication; from the media into primary groups where information is sifted and opinions are accepted or rejected. Sometimes the output from the audio-visual media is used merely as background noise. In Pontinen's (1996) study the young Finns usually ate, talked or rested while they 'watched' television. It was the exception rather than the rule when they switched on to watch a particular programme. These young people had no loyalties to any particular channels or favourite programmes.

Critics of the culture industries have always claimed to take proper account of consumers' choices. Ien Ang (1996) accepts, indeed stresses, that present-day television audiences are plural rather than monolithic, that they have a wider than ever choice of channels and programmes, are active meaning producers, and that their meaning production is deeply embedded in the micro-politics of their everyday lives. Even so, she disputes that this is leading to the disempowerment of the television industry. Rather, she argues that a reconfiguration of the television industry's power is underway: that the 'hard-to-get audience' is being built into production systems, that viewer choice and self-determination are being developed as hegemonic ideas (accepted as simply factually true by everyone), and that this blinds audiences to the limits of the range from which they can choose.

The contrary view, more compatible with the evidence in my opinion, is that most people have their own leisure priorities which are responsive primarily to their economic, gender and age situations rather than implanted by leisure suppliers, that they use radio and television accordingly, and that much the same applies when they are making other purchases. Chapter 6 introduced Robert Stebbins's (1992) concept of 'serious leisure'. In coining this concept Stebbins has drawn attention to how people can become seriously committed, highly knowledgeable, and increasingly skilled in sports, breeding tropical fish, choral singing, wine tasting, astronomy, acting or whatever. Most leisure may be non-serious in Stebbins's sense, but all the indicators suggest that consumers are becoming increasingly knowledgeable and discriminating. Their tastes are not being progressively shaped and homogenized or McDonaldized by advertising and other marketing techniques.

Chris Gratton (1992) has drawn attention to the rise of the skilled consumer. David Darton (1986) has highlighted the same trend in the growth of leisure connoisseurs who have discriminating tastes in wines, visits to stately homes, meals out and weekend breaks. Rather than becoming increasingly passive, consumers are becoming more active and keen to participate in structuring their activities. For example, the holiday purchasing public is becoming more mature. The 'standard package' is becoming less popular (see Henley Centre, 1993). A growing number of holiday-makers want something different, and often welcome the opportunity to construct parts of their itineraries, to pick their own hotels and where to eat. This applies in all product markets.

Hugo van der Poel (1994) has argued that the 'modularization' of leisure supply enables people to pick and mix their own lifestyles from the variety of goods and services that are available. This is not a particularly recent trend. In the 1970s the work on post-war youth cultures at Birmingham University's Centre for Contemporary Cultural Studies (see Chapter 5, p. 121), shattered the view that youth styles were constructed 'up there' then 'sold' to young people. These researchers explained how young people themselves had created teddy boy, mod, rocker, skinhead and other styles by blending products in their own ways, and how the media and other merchandisers had to try to read and follow, but were unable to dictate young people's leisure tastes and activities (see Hall and Jefferson, 1976; Hebdige, 1979).

Supply side pluralism

The reason why consumers are able to make significant choices is that they have many leisure goods and services to choose from. It is true that some leisure business sectors have large, sometimes multinational operators who strenuously attempt to shape consumer taste and demand. This applies in newspapers, broadcasting, holidays, drink manufacture and gambling. However, even the biggest businesses are not monopolies, and even if they were they would not be marketing essentials such as fuel which people have to buy. And the big operators share all their product markets with hosts of smaller competitors. Most restaurants, holiday companies, sports goods suppliers and television programme makers are small businesses. They operate in markets where tastes are fickle, trade is often seasonal, competition intense and the business failure rate high. These businesses do their best to cater to consumers' demands; they cannot afford to try to dictate or even shape people's preferences (see Butler, 1978; Hodgson, 1988; Berrett *et al.*, 1993). Many proprietors of small leisure businesses do it partly from interest. They were often involved in the markets as consumers prior to becoming suppliers, and depend for their success or survival on their inside knowledge, flair and initiative.

Most leisure markets are intricately segmented. Market forces do not standardize tastes. Nor does commerce normally seek out, and cater exclusively for a 'lowest common denominator'. The best recipe for success is normally to identify and cater for a specialist market segment. This applies whether the merchandise is printed matter, housing, sports clothing, prepared meals, holidays or fitness training. Most leisure products can be aimed realistically at only limited sections of the public, which is why enthusiasts with specialist knowledge can often find niches for their small businesses. They know that their viability depends on knowing their customers' preferences, and providing a quality service which is at least a little better or different from everything else on the market (see Roberts *et al.*, 1988). Commerce will cater for all demands, majority and minority, for which customers are willing to pay the market price.

Anyone who is dissatisfied with the range of commercial provisions can look to voluntary associations. These will do anything for which their members have sufficient enthusiasm. This is the engine that drives the voluntary sector, the equivalent to profit in commercial businesses (see Bishop and Hoggett, 1986). Some voluntary organizations cater for their own members' interests. Others aim to meet the needs of other groups. Any medium-sized town has hundreds of voluntary bodies. These are based on hobbies, sports, the arts, and anything else that has enthusiasts. Commercial leisure has not, and cannot wholly absorb the voluntary sector. They are not functional equivalents or alternatives. Each provides its own distinctive experiences and satisfactions. Money is never likely to rule more than a small part of sport. Most teams and clubs are still amateur and exist simply because their players and officers are enthusiastic and want to express their enthusiasm. Voluntary effort not only reduces the costs of participation but delivers special satisfactions.

Then there is the public sector. This will still provide anything that politicians want to provide and pay for. The many reasons why central and local government inevitably become involved in leisure provisions were explained in Chapter 1. For present purposes it is sufficient to note that the public sector adds substantially to the range of leisure provision, and competitive tendering for the management of Britain's local authority facilities is not reducing the scope of the public sector. If they find nothing attractive on offer in any of the 'sectors', people do not have to use any. They can spend their leisure privately and do their own things either in solitude or with friends or family members.

Such is the diversity of leisure supply that no provider, or even one of the main sectors, can stamp a clear pattern on the public's leisure behaviour. Public provision may make a difference at the margin but it cannot force people to do what the providers might prefer. At any rate, public provision cannot be this influential in market economies. Under state socialism, when there was no voluntary sector and commercial provision was minimal, it was possible for the state to make a significant and noticeable difference, though even under communism people at leisure could withdraw into private life in their homes and with their families. In market economies, in countries where people are able to self-organize, it is they who ultimately decide what voluntary associations will do, and which public sector and commercial facilities will be used. Most people, needless to say, pick and mix, and the end result is that the roles of public, voluntary and commercial provision become thoroughly interwoven. Sport teams that run on voluntary effort often play in publicly provided facilities and nearly always use clothing and equipment supplied by commerce.

Explanations of leisure behaviour have to start with types of people – their circumstances, desires and opportunities – not types of provision. It is different when explaining what people do at work, or the education that they receive. People's jobs are organized by businesses that have to meet market demands at market prices. The employers decide how work will be organized, then recruit workers to do the jobs. In school education over 90% of all pupils are in publicly funded establishments and it is state policies that dictate what opportunities will be available. Explanations of the types of education that different groups receive have to begin with the system. Leisure is very different. Exaggerating the force of consumerism and consumer culture is just one example of failing to recognize this.

Strengthening consumer sovereignty

There is another extreme, but possibly more realistic view; that in present-day capitalism the balance of market power has swung decisively in favour of consumers (see Miller, 1995). This argument points out that consumers, who are mostly car drivers, have a variety of places where they can shop and numerous commodities on which they can spend their money. If they stay at home there are many things that they can do or watch. Their consumer choices are registered constantly in supermarket tills and are nowadays fed instantly by computers not just to shelf-fillers but also to the retailers' suppliers. The retailers have to stock what consumers want, at competitive prices, otherwise their market shares will contract, so they must pass on consumers' messages and pressures, ultimately to the producers. According to this argument, it is the pressure of developed world consumers, now the supreme historical force, that is forcing down the incomes of developing world producers and, increasingly, pay levels in much of the developed world also. This adds another dimension to the picture (see Chapter 3) of people's own consumption aspirations subjecting them to heavier pressure in their other roles as workers.

The End of Consumerism?

It is impossible to disagree with social historians such as Gary Cross (1993) who warn against treating modern consumer culture as a permanent fixture. This socio-cultural form has been dominant only since the mid-20th century, and since then only in the Western world, and it is most unlikely to endure eternally. But this is not to say that a demise of consumer-based leisure is imminent.

Marxists have never made such a claim. Rather, they have usually stressed the hegemonic power of consumer capitalism. Their critiques claim no more than that alternative ways of organizing and catering for leisure are possible, and that there are at least signs of 'resistance' in popular cultures.

Box 7.4. The unmanageable consumer.

Gabriel and Lang (1995) claim that the form of consumer culture that became dominant in the mid-20th century is already being eclipsed. They argue that the 1950s and 1960s were the golden age of mass consumption. Its success then is said to have depended upon:

- the ascendancy of Fordist modes of economic organization. Big was best. Mass production created affordable products that could be mass marketed;
- Keynesian economic management, which created secure jobs and incomes, which made people willing to spend and spend, often using credit;
- new types of goods being produced, consumers being immature, and easily persuaded to purchase standardized articles.

Gabriel and Lang argue that subsequently the props on which mass consumption rested have been removed or gravely weakened.

- Jobs and incomes have become less secure.
- People are now experiencing spending fatigue. It has become more difficult to persuade people that yet another purchase will quench their appetites.
- More mature consumers have become fickle and discriminating. Their purchases tend to be opportunistic and spasmodic, and, from the point of view of suppliers, chaotic and unmanageable.
- Businesses are now obliged and are able to respond to new types of consumer demand because new technology has made small beautiful. Small batch production can be as profitable as long runs. Small businesses can afford state of the art technology. Firms are also able to make themselves flexible. This enables them to target market niches rather than the mass, and to respond to constant, unpredictable changes in the volume and types of consumer demand.

The unmanageable consumer

Non-Marxist writers such as Gabriel and Lang (see Box 7.4) have produced more plausible accounts of how consumer cultures are most likely to end, or be transformed. Actually Gabriel and Lang do not predict the end of consumerism so much as its transformation, or perhaps just a gradual evolution from the kind of consumerism that was dominant in the mid-20th century.

Transformed growth

Writers who claim that an end to consumer culture is actually in sight usually do so on the ground that 'things just cannot go on like this indefinitely'. Some of these arguments, critiques of the 'work and spend' society, were encountered in Chapter 3. Martin and Mason (1998) have drawn these arguments together in a critique of social and economic policies which prioritize growth in output, employment and affluence, and allow nothing to stand in the way. They argue that such policies simply create too many problems. Some of these problems have to do with the distribution of paid work and free time. They point to one section of the population having too much paid work, and another section having too little income to use free time to maximum advantage. They claim that current policies and trends are depriving most people of quality free time. They also argue that current trends are placing unsustainable pressure on space and other environmental resources.

Hence their advocacy of transformed growth which would redistribute paid work more evenly across the life span by allowing employees to take more useful 'chunks' of free time (such as sabbaticals) and simultaneously reduce the numbers of people outside the workforce completely. They advocate educating children and young people in purposeful, serious, self-developmental uses of leisure, and locating more leisure facilities in local neighbourhoods in order to reduce travelling.

The weakness in the Martin and Mason scenario is that its achievement seems to depend on a government led re-orientation of education, coupled with public information, opinion forming campaigns to change people's attitudes towards work and free time. It all seems as likely as Marxist prognoses that consumerism will one day be overthrown by working class resistance.

No doubt there will eventually be an end to consumer culture as it is currently known. In the immediate future, however, the most likely trends appear to be an intensification of the consumer processes highlighted by Miller, and the significant but subtle changes in the character of consumption described by Gabriel and Lang. What about the problems? It is most likely that in the first decades of the 21st century, as in the 20th century, people will simply cope.

Chapter 8

Lifestyles and Identities

The word 'lifestyle' may invoke images of Yuppies (young upwardly mobile professional people) and Woopies (well-off older persons). Anyone can join these groups, or so it may appear, by looking the right age, acquiring appropriate clothes and grooming, and either visiting trendy wine bars or taking frequent holidays. All this may seem extremely modern, even postmodern. But in fact only the word lifestyle, or at least its becoming part of everyday discourse, is new. Sub-cultures and contra-cultures have been around for a considerable time. In the 1960s there were mods and hippies. In the inter-war years there were flappers and bohemians. Even the claims currently being made about lifestyles and identities were debated in the early 20th century. Recently these claims have been revamped, principally by exponents of the new sociology of consumption, sometimes drawing on methods developed in cultural studies, and within broader debates about postmodernity (see Chapter 7, pp. 159–164).

Lifestyles

Max Weber, 1864–1920

An examination of leisure and lifestyles really has to begin with Max Weber. This is not merely because this founding father of modern sociology wrote a great deal about styles of life; his statement of the key issues has proved strikingly prescient of current claims about the role of leisure and lifestyles in constructing social positions and identities.

Much of Max Weber's work can only be understood as a debate with Marx; some say the ghost of Marx since it was the Marxism of the late-19th and early 20th centuries rather than Karl Marx's own works with which

Weber was directly engaged. Marx was then regarded as having propagated a mono-dimensional view of social stratification in which other inequalities, of political power for example, were seen as arising from an economic class base. Weber, in contrast, argued that classes, status groups and political parties arose in separate domains of social stratification which were interdependent but irreducible to one another (see Gerth and Wright-Mills, 1948). He defined classes as economic formations. In Weber's view classes were normally formed in market places, in labour markets in the case of the classes to which the working population belonged, whereas status groups could be identified by the degrees of prestige or honour that were attached to their styles of life. Expressed rather crudely, but highlighting the main points, it might be said that Weber believed that their class positions depended on how people earned their money, while their status depended on how they spent it. Weber believed that the Indian caste system illustrated the processes of status group formation, but his main example from his own society (Germany) was the *Junkers*, the counterparts of the English aristocracy, whose way of life had a status which industrialists, whatever their fortunes, could never match.

Weber realized, of course, that class advantages could be converted into status, that some lifestyles required an outlay that set them beyond the means of many classes of people, and that status could be used to advantage in economic markets. He did not claim that class and status were completely independent. However, he insisted that neither was wholly reducible to, arising directly and automatically from, the other. Weber believed that the relative importance of classes and status groups could vary from time to time and place to place. In periods of severe economic dislocation and change he suggested that class divisions and interests might become prominent whereas in periods of relative tranquillity status groups might become more important, as sources of interests and political action for example.

Neo-Weberians

Subsequently even Weber's sociological admirers have not always remained strictly faithful to his original concept of status groups. The term neo-Weberian has often been applied loosely to any non-functionalist, non-Marxist sociology. Market, work and status situations have sometimes been treated as different components of occupational stratification. Such treatment is often described as neo-Weberian, and likewise when occupational classes are defined in terms of their prestige or social standing. American sociologists have commonly used such a concept of occupational class.

The American 'leisure class', about which Thorstein Veblen (1857–1929) wrote caustically, was a status group in the Weberian sense, identified by its conspicuous consumption. Veblen (1925) was critical of this group's role and influence, but he was equally scathing about America's predatory and parasitic capitalist businesses. His hope was that the status of engineers would rise and that there would be a revival of the 'instinct of workmanship'.

In Britain until 1981 the Registrar General's social classes were supposed to be based on occupations' social standing. Since 1981 it has been claimed (by the Registrar General's office) that the classes in fact represent skill levels. Academic sociologists have always been suspicious of this official class scheme because the groupings have never been based on comprehensive and systematic research into occupations' social standing or skill requirements, and in 1994 the Economic and Social Research Council accepted a brief to propose an alternative official scheme. Marxists, needless to say, have consistently rejected all class schemes that group people according to the status or other rewards derived from their jobs instead of their relationships at work – as employers, managers or employees.

Current debates on lifestyles have returned to the issues on which Weber himself wished to focus. Earlier chapters have identified a series of changes that have weakened (though certainly not eradicated) the predictive power of types of employment, sex and age as regards people's uses of leisure. The expansion of leisure itself, the growth in the amounts of time and money that people have available for leisure, has enlarged their capacity to use these resources in a variety of ways. Hence the plausibility of claims that people are now able to use leisure, along with other resources, to construct styles of life that have to be compatible with, but which need not be closely or directly shaped by, their types of jobs, ages or sex roles. Rather than age, sex and occupational groups each having a given *way of life*, some writers claim that members of all these groups now have greater scope to construct preferred *styles of life* which may then become the most significant markers of their social positions and identities. Unlike in the neo-Marxist critique of consumer culture, this lifestyle theory puts 'the people' in charge of their own lives, using commercial and other goods and services for their own creative purposes. This is the attractive side of the lifestyle scenario. It is suggested that through the lifestyles that they construct, people can become almost whoever they want. However, we shall see that there is a gloomier, more threatening side to this theory.

Definitions

Since Weber, numerous definitions of lifestyle have been offered, usually adding precision to Weber's concept but in a variety of different ways. Conceptual dissent is, of course, quite common in the social sciences.

There are alternative definitions of politics, the family and social class, not to mention leisure. Tony Veal (1993) has reviewed the lifestyle literature and discovered numerous definitions – almost as many as there are authors. However, it is possible to identify a more limited number of axes of disagreement. One is whether lifestyles are to be treated as individual or group phenomena. Some authors have used the term to identify the ways of life that express different personalities. For example, extroverts are said to lead extroverted lifestyles, and so on. Other definitions treat lifestyles as properties of groups such as the young upwardly mobile professional people. Another disagreement is about whether lifestyles are to be identified solely in terms of how people behave or whether attitudes and values are to be included. A further disagreement is over what can be included in a lifestyle. Should lifestyles be defined to include everything that people do: their jobs, politics and religious affiliations for example, as well as their family patterns, types of housing and leisure activities?

Conceptual debate does not always indicate disagreement on what exists 'out there' in society and people's everyday lives. Sociological and psychological concepts are tools for looking at and analysing real situations but are not usually intended to mirror everything or anything faithfully. They deliberately focus on particular features and processes, and the definitions selected always depend largely on the analysts' purposes. Social psychologists who are interested in the relationships between personality traits and leisure behaviour are likely to find it useful to define lifestyles as properties of individuals. Sociologists who propose alternative definitions are not necessarily disputing that some leisure differences arise from individual personality factors. For sociologists, however, it will usually be more appropriate to employ the lifestyle concept to identify uses of leisure which tend to 'go together' and characterize a particular social group (Gattas *et al.*, 1987). This is a sensible use of the lifestyle concept, given sociologists' purposes. Sue Glyptis (1981), for instance, has examined the extent to which the leisure of people who visit the countryside regularly differs in additional ways from the lifestyles of the rest of the population. No one has absolute authority to assert how any term must be used. Concepts cannot be appropriated as the private property of sociologists, psychologists or anyone else. However, at some point everyone has to justify their concepts by demonstrating that they lead to significant findings. This is how conceptual disputes are normally resolved.

Tony Veal (1993) has proposed a definition of lifestyle that will accommodate most uses, which is both a strength and a weakness. 'Lifestyle is the distinctive pattern of personal or social behaviour characteristic of an individual or a group.' The problem with this and other common denominator definitions is that it is too bland to serve anyone's purposes. If, as in the current chapter, the issue is whether leisure today enables people to develop lifestyles which then act as bases for their self-concepts and social identities – who they believe they are and how others see them – a particular kind of definition of lifestyle is required.

Lifestyles that confer social identities will necessarily be group phenomena, a 'modern form of status grouping' according to David Chaney (1996, p. 14). Also, the meanings of behaviour, rather than just the behaviour itself, must be included in the definition. And it will be sensible to define lifestyles as being composed of meaningful behaviour over which individuals have some significant short-term choice, such as their leisure time activities. So age, sex and occupation will not be recognized as possible bases for lifestyles in the following discussion, or as part of the definition, though this does not exclude the possibility of people adopting lifestyles that attribute specific meanings to their age, gender or type of employment. The Yuppies, if they exist, identify themselves by the places where they congregate (wine bars for example) and their clothing (smart business attire) from which their types of employment may be (correctly or incorrectly) inferred. The employment itself, however, is insufficient to offer membership of the lifestyle group. Similarly, Woopies (well-off older people) are identified through their high levels of leisure spending and activities; chronological age and wealth alone do not confer the social identity. As Tony Veal (1993) has noted, this is the crucial difference between the *ways of life* that have hitherto arisen among groups surrounded by the distinctive constraints and opportunities associated with their types of employment, ages and sex roles; and the *styles of life* or *lifestyles* that some, but probably not all members of such groups, and perhaps individuals from several such groups, are nowadays said to be able to construct or adopt through their own choices.

Typologies

Seeking a definitive and comprehensive list of lifestyle groups would be setting oneself, or expecting someone else to accomplish, an impossible task. First, if lifestyles are chosen or constructed by the actors, the lifestyle 'map' is likely to be extremely fluid. One would expect some lifestyle groups always to be in the process of formation, some becoming more sharply defined, and others fading away. Any map is unlikely to remain reliable for long. Second, the chances are that different groups will use different combinations of music, clothing, beverages, sport, the arts and so on in constructing their lifestyles. So the lifestyle map at any point in time is likely to be chaotic rather than tidy. Third, it is unlikely that all members of the public will identify themselves, or be identified, with any particular lifestyle groups. When young people have been invited to describe their own sub-cultural affiliations the majority have insisted that they are just 'normal' or 'ordinary' (Brown, 1987; Willis *et al.*, 1988). This seems certain to apply among adults. One would expect lifestyles to appear almost haphazardly, some growing in prominence and others fading on the socio-cultural landscape, with many, probably most, members of the public positioning themselves as detached though possibly interested observers. However, if lifestyles are of increasing importance as sources of social positions and identities one would expect there to be a historical trend towards a growing proportion of the population seeing themselves as belonging to such groups, and a related trend towards the groups becoming more stable, more clearly defined, and more easily identified.

There are lifestyle typologies – several have been developed and used – but in most cases these do not identify specific lifestyles but lifestyle clusters. These may be identified theoretically by analysts who guess that particular processes or divisions are likely to be important, then use the typologies in their own investigations, or hope that someone else will do so. For example, Koch-Weser (1990) has developed one lifestyle typology based on different mixtures of choice and constraint, and a second set of *ad hoc* types: media dominated, intellectual inner-directed, convivial tradition directed, and restrained. However, the value of these schemes has yet to be tested in research.

Other typologies have been constructed and used in market research. For instance, the ACORN typology (a classification of residential neighbourhoods) is a British scheme which groups districts according to their residents' socio-economic standing, types of dwellings and tenure, and age and household composition. This typology can be used to predict the likely demand for, and the use that is likely to be made of, leisure facilities ranging from sports centres to wine bars and arts centres. Tony Veal (1989) describes this and several other market research typologies in the course of advocating status and lifestyle as a pluralist alternative to Marxist class analysis in the investigation of leisure differences.

The socio-economic classifications of UK market researchers that distinguish ABs, C1s, C2s and so on are useful for identifying the motivations and spending patterns of the sections of the population that should be able to afford given products and, therefore, how these should be designed and marketed to maximize their appeal within the relevant groups. These and other typologies have to be judged by whether they serve the intended purposes, but these do not include delineating lifestyle groups with which real people identify. There can be very few people, if any, who regard themselves, or are identified by others as ABs or C2s, or intellectual inner-directed.

Market research has also identified general lifestyle trends such as a recent concern for health and fitness. In the 1950s and 1960s sociologists noted a trend towards home centredness. These trends are not lifestyle groups, though they may serve as a base for such groups' formation. Indeed, the identification of trends in how people live may be a useful first step towards investigating the development and adoption of, and identification with, particular lifestyles within specific sections of a population.

Lifestyles and the Postmodern Identity Problem

The weakening of modern structures

A feature of postmodernity is said, in some versions, to be a weakening of former structures and a blurring of older divisions. The globalization of economic relationships and cultural flows, and the spread of information technology, are usually held to be deeply implicated in these trends (see Featherstone, 1988, 1991; Hughes, 1993). Former fixtures in the local and national social and economic landscapes are said to have become vulnerable. It has been argued that the basis of contemporary economic life is no longer the production of material goods by businesses embedded in specific industries and places but is best conceptualized in terms of 'flows', increasingly of goods and services whose value is primarily symbolic (Lash and Urry, 1994). In these circumstances companies' sites, and workers' occupations and careers even more so, lose their former solidity. Work ceases to be something to be done reliably at particular times and places. Occupations can no longer be relied on (see Chapter 3).

Alongside these trends in economic life it can be argued that sex roles have become more diverse (see Chapter 4), and likewise age roles (see Chapter 5). A consequence, it is claimed, is that neither sex, age nor employment, even in combination, remain good predictors of how people will spend their leisure. Another consequence, it is argued, is that everyday life becomes fragile and the old social markers (age, sex and occupation) cease to supply 'given' identities (Giddens, 1991; Laermans, 1994).

The old, modern social structures created deep divisions, advantages and disadvantages but, it is currently argued, people knew their places and with whom they shared interests. One result could be fierce conflicts, class conflicts for example in industry and politics. Individuals may have been discontent but they knew what they were discontent about, who their allies were, and their enemies. Nowadays it is claimed that these older certainties have disintegrated; that age, sex and occupation no longer give individuals clear positions, interests and identities.

New options

However, compared with the earlier industrial or modern age, many people today have more money, and some have more time at their disposal, and they are surrounded by a plethora of consumer industries supplying goods and services rich with symbolic meanings as well as material uses. This, it has been argued, presents possibilities for people to create identities according to what, where, when and how they consume (Featherstone, 1991). Numerous writers have made this claim in recent years (see Box 8.1). David Chaney (1996) acknowledges that consumption-based lifestyles predate the 20th century, but argues that their development has accelerated with the spread of mass consumption. According to Robert Bocock (1993, p. x), 'Consumption now affects the ways in which people build up, and maintain, a sense of who they are, and who they wish to be. It has become entwined with the processes surrounding the development of an identity.' Alan Warde (1995) has pointed out that attributing such importance to consumption opens the possibility of individuals committing suicide (with their identities) every time they shop. By changing one's consumption style it would appear possible to obliterate an old identity and acquire a replacement, though no one is obliged to be born again repeatedly.

This interpretation of consumer behaviour can offer a more attractive picture of consumption and consumers than the critiques of consumer culture which portray people as passive and manipulated. The identity construction interpretation credits people with the scope and capacity to choose, and portrays them as creative architects of identities that are assembled by making appropriate selections from the 'modules' of consumer goods and services that are available, or which can be generated in the voluntary, commercial or public sectors, or privately.

Box 8.1. Consumption: the new structuring principle.

Pekka Sulkanen (1997), a Finnish social scientist, claims that consumption has now become the main process structuring or patterning people's lives.

She claims that our actions are now governed primarily by a search for happiness, therefore behaviour is more likely to be expressive (of our feelings) rather than instrumental (geared to a more distant goal).

She argues that at present we relate to one another primarily via the 'tribes' that we choose to join by virtue of what, when and where we consume, and that the cohesion of these groups is essentially emotional rather than based on rational calculations of mutual interests.

Old and new 'society of leisure' scenarios

The above account of how leisure has grown in qualitative importance certainly looks more plausible today than earlier forecasts of a society of leisure where leisure values and interests were expected to play a more central and stronger role in people's behaviour in other spheres (Dumazedier, 1967, 1974). The society of leisure forecasts of the 1960s were to materialize as people's leisure values and interests provoked a flight from work, demands for more leisure-like work environments, as leisure became increasingly central in family life, and as people's leisure commitments dictated where they were willing to live, and when and where they were willing to work.

The recent claims of some leisure scholars about the larger and stronger role that leisure now plays in identity formation are paralleled elsewhere in sociology by arguments about occupational classes being replaced by consumption cleavages as sources of collective interests and political partisanship. In political sociology it has been argued that there has been a process of class de-alignment (Crewe, 1989) when class is measured in the conventional way, by occupation, and that the more significant social cleavages increasingly pivot around consumption factors such as housing (whether people own their own homes or rent) and the extent to which they are able to purchase, and actually purchase, health care, education, transport and social security privately as opposed to relying on state services (Saunders, 1990).

These arguments provide a useful reminder that lifestyles are not based on leisure alone. People can also make lifestyle choices by virtue of where they choose to live, their types of dwellings, and the vehicles that they drive. It is also important to recognize that occupations, age and sex are not necessarily, or even likely to be, eliminated totally from the processes of identity formation. No one has made so extreme a claim. Rather, the claim is that the trend is towards these statuses being given meanings by the actors' lifestyles rather than operating as determinants of each group's way of life. It is argued that as lifestyles become more important to people, the socio-cultural value of goods and services, rather than their mundane uses, becomes paramount, and that symbolic capital and a particular kind of communicative competence become vital assets (see Chaney, 1996).

Upper Middle Class Lifestyles

Pierre Bourdieu

There have been several recent studies of the development of new lifestyles within sections of the upper middle classes, or the service class as these strata are now usually described by sociologists. Most of these investigations have acknowledged their debt to Pierre Bourdieu, a French sociologist who has studied the audience for high culture in France and related the findings to his broader interest in social stratification. Bourdieu (1984) contends that, in France, cultural production and consumption are part of the class structure and serve simultaneously partly to conceal and partly to reinforce and reproduce economic and political class relationships (see also Bourdieu and Darbel, 1997). Bourdieu recognizes and indeed stresses that there is not a perfect fit between France's economic, political and cultural elites. His argument is that their manifest separation is important for the reproduction of them all. Despite their manifest separateness, Bourdieu shows that the audience for high culture – visitors to classical concerts, art galleries and so on – is heavily skewed with the economically advantaged and politically powerful grossly over-represented. He argues that cultural taste is one way in which the members of privileged strata can instantly recognize one another and their children, and exclude other sections of the population from their circles. Moreover, when seen in public the members of these strata are likely to be displaying their cultural tastes which thereby become prominent parts of their social identities. Those concerned are able to regard themselves, and hopefully be regarded by others, as persons of superior tastes, sensibilities and abilities, rather than self-interested wielders of economic and political power. Exploitative class relationships are thereby given a more acceptable veneer which legitimizes, consolidates and permits the reproduction and perpetuation of these inequalities.

As a general theory of class structure and relationships Bourdieu's arguments are plainly inadequate. Too few members of the upper middle classes are sufficiently versed in high culture for this to be much use in enabling them to recognize each other, or in seeking wider esteem for their social strata. Chapter 3 explained that over time class-based cultures and ways of life have weakened. The upper strata have tended to become leisure omnivores (at least on a collective level) who have sufficient money to nurture and indulge their typically wide-ranging tastes. Moreover, Bourdieu's explanation of how social classes are reproduced inter-generationally cannot be easily harmonized with the volumes of upward and downward mobility that are known to occur, or the processes whereby mobility is accomplished (see Goldthorpe, 1996). However, Bourdieu's ideas can be applied to status group formation within sections of the upper middle classes which, in fact, is how the ideas have typically been used.

Super yachting and the super rich

George Hughes (1993) has offered a Bourdieu inspired interpretation of super yachting by the super rich. Why do they spend so much money on vessels that they rarely sail personally, which cannot offer the comfort of five star hotels but largely protect the passengers from, rather than expose them to, the challenges of the ocean elements? Hughes's explanation is that the actors are attracted by the social identities that super yachting confers. The vessels advertise their owners' wealth while providing them with a public image as people who are basically interested in pleasure and challenge with which the population at large can sympathize. Most of the wealth of the super yacht owners may be employing workers on low wages in dangerous or unhealthy conditions, but on the public stage the super rich are able to present, and enjoy, alternative identities.

Life on The Heath

Derek Wynne (1990, 1998) has studied the construction of social position among the rather less prosperous, but still quite well-to-do residents on The Heath, an upmarket private housing development in north-west England. The prices of the dwellings excluded all but the solid middle class. All the households had members who were professionals, managers or self-employed. To outsiders with ordinary jobs, The Heath's residents would have appeared uniformly privileged. However, Derek Wynne explains how different groups of residents developed different lifestyles, involving different uses of neighbourhood and other facilities, and in this way The Heath became divided into different types of people. One lifestyle group was composed mainly of highly educated (graduate) professionals who played sport, went to avant-garde theatre productions, and selected à la carte or self-constructed holidays. Their lifestyles, and social identities, were quite different from those of a group composed mainly of managers and the self-employed who participated in a drinking culture, bought expensive home furnishings, meals in steak houses, tickets to big shows in big cities, and packaged holidays. Wynne claims that nowadays middle class social identities are typically rooted not in the people's work, but in their types of consumption and leisure practices, that these processes are helping to fragment the middle classes and are outdating the view of such groups as uniformly conservative, socially and politically.

Professionals and managers

A similar distinction between the lifestyles of professionals and managers was found by Mike Savage and his colleagues (1992) in their analysis of the spending and activity patterns of the upper middle classes, using data from large-scale market research surveys. Savage and his colleagues actually distinguished two lifestyle groups among the highly educated professionals. One pursued health and fitness through their choices of food and regular exercise. Another was identifiable by its cultural activities – visits to the theatre and suchlike – and a fondness for entertaining friends to dinner and drinking wine. The lifestyles of the managers in this study were relatively undistinguished. It is not an entirely unfair caricature to say that their tastes were not very different from manual workers' except that the managers were able to spend more on them.

Alan Warde's (1995) analysis of food consumption in Britain's upper middle classes reached a similar conclusion. The highly educated professionals had food purchasing patterns that clearly distinguished them from manual workers: far less bread, potatoes and beer, and more fresh fruit, wine and salads. Once again, however, the managers' diets seemed best described as 'undistinguished'.

Other recent studies of Britain's middle classes have suggested, or at least implied, that sections of these groups recognize each other, and identify themselves, as much by their out-of-work lifestyles as by their types of employment. For example, Butler and Savage's (1995) edited book on *Social Change and the Middle Classes* contains just three chapters on the middle classes' employment but six which deal with their housing, uses of the countryside, food consumption, and urban gentrification (see also Thrift, 1989). This book argues that social classes need not be formed in workplaces and occupational associations but can be constructed in the districts where their members reside or which they visit for leisure in their green wellies or Range Rovers, for example, thereby signalling their group membership.

These arguments credit leisure, or consumption as most of the writers prefer to describe it, with a new importance. When we play sport, go out for a drink, on holidays, or day trips it is claimed that we are no longer just relaxing, letting off steam or otherwise 'being' ourselves so much as creating selves, forming ourselves into social groups with which we identify and which give us public identities. Some believe that it is in such groups that we are now most likely to be mobilized for political action to defend 'our' countryside against motorway construction, or to have motorways routed so that 'our' quiet villages remain uncongested, for example.

Postmodern Leisure

As already indicated, there is a gloomy side to theories of people turning to leisure-based lifestyles to construct self-concepts and social identities. Writers on postmodernity have recognized that the forces that are destabilizing employment, gender and age roles are unlikely to leave leisure unscathed. After all, leisure is affected by the same new technologies, and the same globalization, that are transforming economic life. In a world where everything else changes rapidly leisure is unlikely to be an oasis of stability. There is a threatening dimension to forecasts of people increasingly turning to leisure and consumption to define who they are and their places in society. People may turn to leisure because it is available, and pliable. Lifestyle experimentation can appear attractive, even fun. Yet people may find the lifestyles, and the social positions that they thereby create for themselves, disintegrating almost as quickly as they are constructed.

There is a substantial but maybe healthy measure of conscious exaggeration in most portraits of a postmodern condition. Writers focus on, and extrapolate specific trends and tendencies, and may thereby lose a sense of proportion. This can still be useful for highlighting specific developments provided no one forgets the distortions that are being introduced.

This certainly applies when characteristics of postmodern leisure are listed. These should not be viewed as even near descriptions of what present-day leisure is actually like so much as characteristics that can be expected to grow in prominence alongside a broader postmodern condition. With this proviso, three features of postmodern leisure have been identified by several writers, and all of these features are claimed to be becoming increasingly evident (Denzin, 1991; Laermans, 1994; Maffesoli, 1994; Rojek, 1995).

Leisure loses its former grounding in a surrounding material reality

First, it is claimed that leisure places and experiences are becoming decontextualized. For example, it has become possible to enjoy tropical holidays in cold climate countries, while theme parks can create safari conditions, or the Wild West, in any place. The Internet allows us to communicate with others irrespective of their locations (provided they have the technology). Virtual reality technology may soon allow people to experience any situations irrespective of their physical locations. Hence the alleged appearance of the 'post-tourist' who finds the actual travel tiresome and grows weary of being shown especially arranged spectacles, and relies more on TV travel programmes and suchlike to reveal places as they really are, and what the live tourist will be, and will not be, allowed to see. In a similar way so-called 'post-fans' may follow sports teams primarily 'on the box' whose presentations may be regarded as the real thing for which attending matches is a poor substitute. Distinctions between the real, imitation and fiction may thereby break down. Disney parks appear to be treated as just as real places to visit as the Florida Everglades and the Champs-Elysées. Soap opera dramas and characters are reported in newspapers and discussed as if they were real. Another manifestation of the loss of 'grounding' is said to be that the (fictional) public images of celebrities, living and dead, exert a powerful hold on people's consciousness.

Leisure becomes chronically unstable

Second, in postmodern conditions it is claimed that cultural commodities and their meanings change constantly. Everything speeds up and leisure environments become discontinuous and fragmented. This, it is said, leaves people feeling insecure, and that nothing can be relied on. In these circumstances people may see no reason to assume that things must remain, or will remain, as they are. They are likely to feel at risk and to become sensitive to the harm that they might suffer from the foods that they eat and the environments in which they work and play.

It has been suggested that people can react to such conditions in several ways. One is social and emotional disengagement. Another is to escape into fantasy; to abandon oneself to postmodern experience. Another is to seek security and belonging in the 'new tribes' or 'proto-communities' that develop around cultural products such as sports teams and musical genres.

The above types of leisure do not appear to be promising bases from which to construct social positions and identities. Building and maintaining an identity from such materials seems likely to require considerable and continuing ingenuity and effort. And even such effort is likely to see its achievements disintegrate repeatedly. This is how the leisure-based lifestyle scenario can be portrayed as a threatening prospect. Strong attachments may be formed to particular sounds and musicians, and intense fellowship may develop among the followers, only for the genres to disappear and be replaced by new commodities. Identities based on postmodern leisure are likely to be fragmented and chronically unstable. Indeed, the postmodern citizen at leisure has been portrayed as perpetually anxious, insecure, restless and unable to settle; a voyeur adrift in a sea of symbols (Denzin, 1991).

Ulrich Beck and Elisabeth Beck-Gernsheim (1995) have described a similar situation as individualization and the weakening of older structures enable men and women to relate on more equal terms and seek in each other love and warmth in an otherwise uncaring world. Meanwhile, it is argued, couples' individual commitments, in the labour market for example, are pulling them apart like never before.

As already mentioned, there is a measure of conscious exaggeration in all this. No one really believes that consumers are distressed and anxious every time they make a purchase, fearful that the lifestyle for which they are opting will disappear or be degraded. As Alan Warde (1994) has pointed out, consumers have an array of coping strategies such as relying on the judgements of close friends. The crucial point is that if older social structures are being subject to flexbilization, leisure may become an arena in which individuals try to restore order and stability to their lives. If so, these efforts are liable to be undermined by the same trends and conditions that provoked the attempts.

Rattansi and Phoenix (1997) have queried whether people today possess stable, core identities. Until recently, researchers have usually assumed that people do possess enduring conceptions of themselves, attitudes and values. So they have enquired, for example, whether leisure might be replacing work as the principal or normal source of these identities. Rattansi and Phoenix invite us to entertain the possibility that all roles and identities have become variable and fluid. In this case consumer-based roles will supply just some of individuals' temporary, often situationally specific, identities.

Everyday life is aestheticized

Third, the collapse of modern structures is said to lead to people's concerns for beauty, appearance and pleasurable experience breaking out of leisure time and infiltrating their working lives, family lives, shopping expeditions and so forth (Lash and Urry, 1994). This is one of the trends that is supposed to favour experiential leisure concepts. People become sensitive to the design of the furniture in their homes and offices, and the cans and bottles in which they purchase all manner of commodities. This reminds us once again that in constructing lifestyle-based social positions and identities people are unlikely to confine themselves to, or even operate mainly with, conventionally classified leisure goods and services. And the threatening side of lifestyle-based identities is once again highlighted. David Chaney (1996) argues that as lifestyles shape our selves and sensibilities, and as everyday life is aestheticized, we find constant reasons to worry about our own surface appearances, and about the meaning of other people's appearances, and constantly feel that our identities are at risk.

One might protest that the above sketches of postmodern leisure stray far beyond the bounds of useful exaggeration and bear little resemblance to most people's lives; that few people spend much time in virtual reality situations or theme parks, that most people know the difference between the real and the fictional, and, in any case, that none of the features attributed to postmodern leisure are really novel. Earlier generations escaped into fantasy at circuses and fairgrounds, in music halls and cinemas. The appeal of Jesus shows that fascination with the dead predates modernity. The crucial points, however, are that if our social identities and self-concepts are becoming lifestyle based, the stakes will be higher than in the past due to the absence of other reliable foundations; and if a postmodern condition really is undermining other bases for secure social positions and identities, the same trends can be expected to wreak similar devastation across leisure.

How Widespread? How New?

Actually, the evidence that lifestyles are becoming more important in the manner suggested above is not totally compelling. Perhaps different sections of the upper middle classes, even those who live side by side, differ in their leisure tastes and habits and thereby distinguish their social positions and give themselves distinctive identities. Is this new, and is it saying something really important about our present social state? Maybe some households and individuals are constructing preferred identities, and becoming known as 'that type of person', by purchasing super yachts, gentrifying parts of the inner-cities, enjoying high culture, or wine, or weekends in the countryside, or playing sport. But how many people are into this lifestyle scene? And are they pioneering new social processes? Young people have a history of forming youth sub-cultures that provide distinctive identities but, as mentioned earlier, when representative samples have been questioned about their own sub-cultural affiliations the majority have continued to insist that they are just ordinary. Surely the same applies to adults, and has always done so.

Once again, Graham Murdock's (1994) interventions are important. He has stressed that a minority, but a substantial minority of the present-day population, remains almost completely excluded from the consumer society. They are precluded by their poverty from developing identities by purchasing appropriate clothing and being seen in the right places. Sheila Scraton (1994) has argued a similar case for the situations of women.

Lunt and Livingstone (1992) conducted a detailed study of the shopping practices of 279 adults in southern England. This sample was questioned about their attitudes towards saving, borrowing, debt and possessions, and their ideas about what was a luxury and what was a necessity. Shopping practices were also explored in detail – where and when they shopped, and what they bought. From their analysis of this evidence Lunt and Livingstone concluded that no more than a quarter of the sample could be described as lifestyle shoppers. The rest were making nearly all their purchases in customary ways to meet their needs.

The same applied to the Glasgow households on post-war private and council housing estates who were questioned by Madigan and Munro (1996) about their home furnishings. Hardly any of these consumers had gone shopping for a style that would give their homes, and themselves, a distinctive identity. Most had built up their stocks of furniture over the years and possessed a mixture of styles. When making new purchases they were influenced primarily by price and use values. Rather than seeking homes that would align them with wider social groups, this sample invoked family values. Above all else they wanted homes that felt comfortable yet looked tidy.

Colin Campbell (1995) has ridiculed some of the more extravagant claims about the social significance of present-day consumption. He accuses writers of attributing symbolic meanings to behaviour which is really just instrumental. Campbell claims that most purchases do not have definite meanings. Buying flowers for a woman is an exception. What, though, he asks, is the significance of wearing blue jeans to go shopping? Campbell shows that most members of the public are hopeless even at judging the prices of most things, let alone reading some agreed symbolic significance into their acquisition and possession.

Prototypical Groups?

New middle class positions

Writers who believe that lifestyles are becoming more important as sources of social positions and identities do not need to claim that everyone, or even most people are involved, yet. They recognize that it is the relatively well-off who are the best able to make lifestyle choices. They also acknowledge that, up to now, such conduct has become common only within specific sections of the relatively well-off, namely, according to Scott Lash (1990), those in new occupations or types of employment that have expanded or changed markedly in recent years and confer no customary identities with which the new role players can, or wish to be identified. Jobs in finance, the law, and especially the new communications media are said to fall into this category. As will the student role, given the recent expansion of higher education which has denied graduates the certainty, or even the likelihood, of subsequent 'glittering prizes'. It is argued that lifestyle experimentation is most likely by groups who live and work in symbolically rich environments, the media and post-compulsory education being obvious cases. The claim is not that everyone now has a lifestyle-based identity but that the groups with the need and the means to construct positions and identities in this way are growing as a result of broader social, economic and cultural trends.

Are these groups really growing in size? If they are still confined to a minority, is there any reason why we should believe that lifestyle-based positions and identities are becoming more common and important? Is the need sufficient to provoke this response? And are lifestyle-based identities the only conceivable solution to the need? Is the need itself new? After all, Max Weber was aware of the existence of status groups. 19th century professional gentlemen were recognized by their dress and social conduct. And what about the stream of youth cultures that began before there was even thought of post-industrialism, post-Fordism and postmodernism?

Those who believe that we are now witnessing something new can argue that in the past, despite Weber's claims about irreducibility, lifestyles normally expressed identities with other, more secure bases. Using the terminology recommended above, in the past there were ways of life rather than lifestyles. The aristocracy had their estates as well as house parties. The professional gentleman had a profession.

Youth cultures

In the past youth cultures developed along class lines, with clearly demarcated gender roles (see Chapter 5). Middle class youth celebrated their privileged status. Working class youth also celebrated their class characteristics: their toughness, lack of respect for authority, and so on (Willis, 1977). In the 1960s mods could celebrate their upward mobility (Cohen, 1972). Other working class youth cultures could be interpreted as attempts to recover identities that were at risk amid industrial and urban change (Cohen, 1976).

The claim is that since the 1970s or thereabouts neighbourhood communities have disintegrated and even occupations that have been entered, not to mention longer-term career prospects, are now too insecure to act as bases for stable identities. Likewise, it is claimed, greater variations among men, women, and in all age groups in terms of their education, occupational and domestic situations have stripped these markers of their former identity-conferring capacity. Hence the claim that young people increasingly construct identity-conferring positions through their lifestyle choices.

Gendered identities

Frank Mort (1996) has claimed that when commerce targeted the young affluent male consumer in the 1950s the market was pre-structured in that the young males in question already possessed core identities, and purchasing predispositions, arising from the neighbourhoods where they lived, the schools they attended, the jobs they entered, and the peer groups in which they thereby became involved. By the 1980s, in contrast, Mort argues that advertisers and the media more generally were playing more active roles in actually constructing images of masculinity which, above all else, were associated with lifestyles based on particular sorts of clothing, leisure activities and places.

Contemporary youth scenes

Young people, since they have still to acquire adult identities, are commonly cited as the vanguard group *par excellence* in respect of the above trends, and Chapter 5 explained that present-day youth cultures are different in several important respects from their counterparts of the 1950s and 1960s. It is but a short step to conclude that nowadays young people typically develop leisure-based lifestyles and identities (see Box 8.2).

The functions of night-life

Robert Hollands (1995) argues that for young people the night out has now ceased to be a *rite de passage* in a brief process of transition between dependence on one's family of origin and establishing an independent

Box 8.2. Young people's spending decisions, social identities and group affiliations.

S. Miles and his colleagues (1998) conducted a questionnaire study among 285 A-level students which reveals some of the complexities in the processes whereby young people may develop consumption-based identities.

The young people's most frequent purchases were music, clothing and footwear. They were not slaves of fashion. Spending was usually planned carefully in advance. Both price and brand reputation were considered important.

The young consumers were not usually trying to be the same as other members of their peer groups. Rather, they wanted to establish individual identities, to 'stick out'. Yet fashion was important. The young people were sensitive to the opinions of their friends. They wanted to 'stick out' in socially acceptable ways.

household, (which used to depend principally on finding a mate) and has now become a prolonged socializing ritual. Hollands argues that city centres have become places whose main function is to supply the night-life which attracts males and females from teenage upwards, from all social backgrounds. He claims that on their nights out young people are able to establish identities by virtue of the crowds with which they congregate, the pubs and clubs that they visit, the types of music played in these places, and the appearances that individuals cultivate. Rather than expressing, celebrating or rebelling against social positions with other bases, Hollands argues that youth cultures now create social positions and identities which may then accord meaning to the individuals' ages, sex roles, educational and occupational statuses.

Such a claim is difficult to test. Measuring self-concepts is complicated. It is impossible to go back in time to replicate the measurements and discover whether the role of young people's leisure activities has really changed. But we can assess the plausibility of the role claimed for present-day lifestyles. As we have already seen, suggestions that leisure can now do what occupations, age and gender no longer can have difficulty in surviving this plausibility test. We shall see shortly that this applies even to present-day youth cultures about which it has been possible to assemble the crucial evidence.

Persistent Boundaries

The fact that age, sex and occupation are now structured less rigidly than in the past will not prevent people continuing to use them to construct a sense of who they are. Much of the new flexibility is in fact within long-standing social divisions. Men and women may now be able to choose from among wider arrays of gender roles, but most of their options remain either gendered or are capable of being gendered, and sex itself is certainly not being eliminated as a social marker. The same applies to age. The period between childhood and adulthood may have been prolonged, and those in transit may now have a wide choice of cultural styles, but these are generally considered suitable for people of their age. The typical age of retirement may have been lowered but teenagers cannot pass as well-off older people. Nor are 50 year olds acceptable in youth scenes provided they only dress appropriately and like the right music: the majority crowds stubbornly insist on viewing older figures as intruders from their parents' generation. Socio-economic classes, if not specific occupations, still have fairly stable memberships. There may be more mobility between occupations than in the past but people still tend to stick at the same levels. Most people appear well aware of their levels and those of other people in their social worlds. Whether they describe the people who occupy the same levels as social classes is beside the point. People know where they themselves and others are located in the socio-economic hierarchy, and these levels continue to operate as powerful social markers and sources of identity (see Marshall *et al.*, 1988).

All the evidence continues to show that lifestyles tend to form within social classes and respect their boundaries, as applies also with age and sex divisions. This was the case with all the examples of upper middle class lifestyles that were discussed earlier. None of these lifestyles was conferring identities that overrode class divisions. The lifestyles may have added lateral distinctions, and fine detail, but were operating within, rather than as alternatives to, more conventional, and still more stable social markers. There is in fact more evidence of the middle classes imposing their meanings on the countryside as a desirable place to live or visit, and of places needing to image themselves (Urry, 1995a, 1995b), than of the new middle classes actually being formed through their images and practices of rural living (Cloke *et al.*, 1995).

The population today is too mobile for places, or even broad types of housing tenure, to provide the kind of long-term interests and identities supplied by occupational careers, and to form classes that reproduce themselves. There is more evidence of occupational class determining where people live, and their types of housing, than the latter operating as bases for class formation (Wait, 1996).

None of the examples of lifestyle-based middle class formation points to the obliteration of more familiar social class groupings. Marxists have always claimed, contrary to Weber, that status groups are normally formed upon or within, and *are* in fact reducible to, class divisions. Even neo-Weberian class theorists have conceded this point in so far as they have treated status as an attribute of occupations or as one component of social class. The postmodernist attempt to assert lifestyles as alternative sources of social positions and identities is not justified even by the limited lifestyle evidence that its proponents present. Bourdieu himself has always insisted, with justification, that cultural consumption be treated as just one part of economic and political stratification.

The main differences in leisure behaviour are still by social class, age and sex. The principal conclusions from conventional leisure research are still proving robust. Sex, age and socio-economic status continue to be related to clear leisure differences; clearer than the differences between the intra-class lifestyle groups that have been identified in existing research. For example, Alan Warde (1995) has shown that although there are differences between the professional and managerial groups, the clearest contrasts in food consumption are still between the middle class and the working class, and the differences between these strata did not narrow between the 1960s and 1980s. Indeed, some social class leisure differences have become even more pronounced. As Chapter 3 explained, income has become the most prominent source of inter-occupational leisure differences, while some of the distinctive leisure patterns associated with specific occupations and work situations in the past have become blurred. Some leisure interests and activities have become more exclusive than previously. For example, Knulst (1992) has shown that in the Netherlands the live audience for high culture has become more exclusively upper middle class. This has been due to price effects, and to the ability of the non-core audience to satisfy its tastes from the radio and recorded music.

Young people's lifestyles reassessed

Lifestyles themselves, in so far as these embellish and add detail to individuals' positions and identities, appear to be constructed on the basis of educational attainments, preferred schooling for one's children, types of housing and transport, as well as leisure tastes and activities. Leisure activities, especially allegedly postmodern ones, will be too ephemeral to be used as bases for secure identities. This applies even among young people. Positioned as they are between childhood and adulthood, many young people have no unambiguous social class locations, and some former social class differences in their leisure behaviour have become less distinct than in previous decades (see Roberts and Parsell, 1994). In the study of night-life in Newcastle upon Tyne that was referred to earlier, Robert Hollands (1995) found that students and local young people were equally involved in city centre scenes. However, he also noted that these two groups tended to go to different clubs and pubs, and that the students were especially keen on venues with live music. In this instance, as in other research, the main tendencies towards the formation of lifestyle groups were occurring within social classes, and within a particular age group.

This was equally the case among the much larger sample of young people in the longitudinal study in Scotland conducted by Leo Hendry and his colleagues (1993). They were able to identify a number of lifestyle groups but all of these tended to draw their members from specific social classes and did not override and obliterate social class divisions. When individuals' lifestyles changed over time their movements were nearly always among lifestyle groups characteristic of their social classes. David Chaney (1996) is surely correct in arguing that, while lifestyles may be growing more important, they do not replace but usually emerge from and within, and coexist with, ways of life with more traditional bases.

As yet there is absolutely no evidence, only assertion and conjecture, about young people's (and adults') leisure activities acting as a principal base for their self-concepts and social identities. John Bynner and Sheena Ashford (1992) investigated this possibility in the evidence from the Economic and Social Research Council's 16–19 Initiative, a set of longitudinal investigations among representative samples of young people from four parts of Britain: Kirkcaldy, Liverpool, Sheffield and Swindon (see Banks *et al.*, 1992). This research found that its samples' social and political attitudes did not cohere neatly into a limited number of clearly defined orientations. However, socio-political attitudes were more strongly related to the young people's family backgrounds, sex, past educational attainments, and experiences in education and the labour market after age 16 than to their leisure activities. In fact there were hardly any clear connections with leisure patterns. This suggests that their leisure activities were not among the principal sources of the young people's ideas about the kinds of adults that they would become and where their long-term interests lay. Class trajectories and gender roles rather than leisure were continuing to play this role even if they were unable to supply as well defined and powerful collective identities as in the past. Bynner and Ashford found that high levels of participation in uses of leisure that brought them into contact with peers enhanced the young people's self-confidence and gave them positive self-images. As previous chapters have explained, it is well-known that these kinds of leisure activity improve people's quality of life. Leisure can make people more satisfied with whatever they are but leisure interests are too transient or ephemeral to tell people who they are.

Most people in all age groups still appear to regard their leisure as just play. Robert Stebbins's 'serious leisure' may be mistitled (Stebbins, 1992). People may become heavily involved in, and committed to particular leisure interests, activities and groups yet continue to realize that it is all basically a game, incapable of producing the serious rewards and penalties that can be incurred at work and in family life. Wendy Eygendaal's (1992) study in the Netherlands found that the Death Metal fans who attended the concerts, bought the music and looked the part in their leisure time insisted, when questioned, that all this was really just acting, simply weekend play, rather than expressing their authentic selves. In the early 1960s Ralph Turner (1964) found that young Americans' commitments to youth cultures were nearly always superficial, and there is absolutely no evidence that this has changed. Leisure still performs its modern, now traditional functions and can continue to do this because of the absence of serious, wider, long-term consequences.

It is also relevant that if one of the traditional markers declines in value, it can be replaced by others. For example, if occupations cease to offer clear and stable collective identities, people can turn to sex or indeed to regional, national, religious and ethnic divisions, and this is much in evidence in our present-day world.

National Identities

A universal base for identities

Identities are complex. They contain layers and compartments that are activated or remain dormant depending on the situations and the prompts. No father is likely to feel paternal constantly. Sometimes, for example, he will be a plumber, a senior citizen or a squash player. We all have leisure identities. This is not at issue. The issue is which identities endure and generate conceptions of interests which take precedence when choices have to be made, and which underwrite our strongest and deepest attachments to other people.

Leisure activities are sources of, and simultaneously allow us to express other layers and compartments of our identities. Playing sport may express a person's healthiness and masculinity. However, a type of identity that is expressed and affirmed in most uses of leisure is ethnicity or nationality. These are expressed in our languages and dialects, and through most leisure interests. The films and television programmes that we watch, and the newspapers that we read, express and confirm a sense of sharing interests with 'our' people. This is not to say that we are usually attempting to demonstrate superiority or express antagonism, though we are unlikely to feel comfortable if we witness our own culture being disparaged. Few people can be indifferent to whether their's is the majority or a minority national or ethnic culture in the place where they live. The leisure activities that go on around us contribute to a sense of belonging to or being apart from the surrounding people and their culture. In any multi-national or multi-ethnic society these divisions and the associated feelings will be expressed constantly in leisure. It should be no surprise, therefore, that although sports are played by both the Irish and British in Northern Ireland these groups often play different sports, and if they play the same sports they are likely to play in different places (see Sugden and Bairner, 1986; Knox, 1987; Roberts *et al.*, 1989a). Within a global trans-historical context such behaviour is normal.

Since the Second World War the cultural norm in Western countries, among their intellectuals at any rate, has been to regard ethnic and national divisions as rather unfortunate. Deviating from this norm has invited charges of racism and even fascism. Good citizenship, we have taught each other, means ignoring all indications of ethnicity and nationality in most situations except possibly to check that we ourselves are not guilty of oppressive practices. In Britain, over the last 40 years, there has been a tendency to treat the country's race relations as 'a problem' arising from Britain's particular colonial history and its cultural legacy, and research into leisure and ethnic divisions has usually been conducted within this race relations paradigm. 'The issue' has been the treatment of minorities and their cultures. The national majority has entered the picture only in so far as it has been racist or otherwise prejudiced, engaging in overt or covert, direct or indirect forms of discrimination.

Actually the division of humanity into various peoples has been constant throughout history. All peoples have possessed an awareness of their ethnic or national identities, not just minorities, or bigots in majority populations. Nationality and ethnicity answer questions to which everyone appears to want replies. We all seem to want to know where we, as individuals, are from: where we were born, who our parents are or were, and the places where they were reared. We also want to know which broader categories of people we belong to and their places in history and the current world. Such categories of people usually recognize each other through some combination of physical appearance, language, dress, religion and way of life. They may believe that they all originate in a common place, from common parents or a deity. Throughout history aliens have normally been viewed with fear or condescension, and often a mixture of the two. Most peoples have felt it important to be able to defend themselves economically, culturally and militarily when necessary. No people has ever been indifferent to the prospect of being transformed from the majority into a minority group, especially in what the people have considered to be their homeland. What applies to the present-day British and Irish in Northern Ireland, Russians in CIS countries outside the Russian Federation, Serbs in Bosnia, Scots who resent rule from Westminster, and English who fear rule from Brussels has many historical antecedents.

National identities are often complicated. They alone can contain various layers and compartments. The same individual can be a citizen of the United Kingdom (and determined to maintain the Union), British, English and European also, just as Armenians, Georgians, Ukrainians and Russians were also Soviet citizens until 1991. And the world's political and cultural maps have never been etched in stone: they have always been 'negotiable', though national identities usually prove more difficult to change than political boundaries. People can and do replace old with new national identities but these processes normally take generations and centuries. Majority groups may seek to assimilate minorities though in practice they have been more likely to seek their subordination and respect. Minorities have been more likely to seek equal rights for their cultures than wholesale assimilation. Ethnic and national identities have been preserved for centuries in families and villages even when they have been oppressed by the authorities. In time the British, Germans, French and so on may come to regard themselves, first and foremost, as Europeans. Such a supra-national identity is most likely to be adopted if it does not require older affinities to be jettisoned instantly. These are certain to be preserved for many generations especially in situations where individuals are most able to 'be themselves', as during their leisure. This sphere of life is, and has always been pervaded by national characteristics. It normally expresses and reinforces the ethnic or national identities of minority and majority populations, and does so irrespective of whether the people involved are racially prejudiced or anti-oppressive.

Race and leisure in Britain

Leisure research in Britain has paid little heed to how the majority population's styles of life express its nationalities; but the equivalent behaviour of ethnic minorities has been explored. Minorities from Islamic and Asian backgrounds have typically entered Britain with strong cultures already shaped over centuries which most of the migrants have valued and have wanted to preserve rather than discard. They have brought into Britain their own conceptions of what it means to relax and have fun. Despite its appearances to native Britons, Islamic leisure does not feel austere to the participants when, for example, they respect the ban on alcohol and segregate the sexes (Ibrahim, 1982). However, Michaelson (1979) has described how East African Hindus living in London and Leicester in the 1970s disapproved of the 'frivolous' uses of leisure by which they were surrounded in their new country, and how they preferred to spend their free time in their own community associations, participating in activities, listening to music and songs, and watching films and dances which maintained their cultural bonds and identities. The children of such first generation immigrants have been more aware of the attractions of the Western way of life but, up to now at any rate, they have rarely totally discarded their minority cultures. Fleming's (1995) research among Asian and white pupils at a north London comprehensive school found that the ethnic minority's under-representation in sport was entirely due to the young people's own preferences rather than racial barriers erected by the wider society. Lovell (1991), and Carrington and his colleagues (1987), have also found that young Asians, especially the females, have little interest in sport, pubs and wine bars. This has been partly due to parental restrictions but to a large extent the young women have internalized the relevant values. They have wanted to preserve their 'own' culture, and their under-representation in much normal British leisure has been primarily self-imposed.

A second major theme in British race and leisure research has been the extent and impact of racism. Asian settlers may have been somewhat less exposed than blacks from Afro-Caribbean backgrounds. The former's own cultures, their family and recreational patterns knit by language and religion, have helped to keep them apart from the host society. Afro-Caribbean settlers have not imported traditional cultures of similar strength. In most cases their original ethnic cultures were largely annihilated under colonialism and slavery. Their pre-migration ways of life were not traditional and valued but developed under conditions of oppression. Most of these migrants were seeking to escape from these ways of life rather than aiming to rebuild and preserve them in Britain. In some cases they entered Britain seeking and expecting rapid assimilation which proved impossible for a variety of reasons including discrimination and racism. Asian settlers have not been immune. Fleming's (1995) research in the north London comprehensive school showed that although the Asians' under-representation in sport was due basically to their own preferences, the sub-cultures which they formed as 'victims', 'straights' or 'street kids', for example, were alternative ways of coping with racism. Even so, the Afro-Caribbean settlers who have become involved in sport, drinking and music cultures have been the more exposed to Britain's race relations. In virtually all sports black players have routinely faced racial abuse from other players and spectators. In professional sport, football for example, blacks have been most strongly represented in the Premier League suggesting that they have needed to be supergrade in order to gain acceptance. When they have been present they have tended to play in peripheral rather than central team positions, and they remain very much under-represented among coaches, managers and officials (Maguire, 1991). British blacks are not under-represented in most 'normal' forms of British leisure but they tend to congregate in their own districts, teams, clubs and musical groups. A principal reason for this segregation has been fear of racial abuse, verbal and physical. Young blacks in Liverpool know that their safety is at risk when visiting city centre clubs. They also fear victimization by the police. These fears are shared by many young whites but blacks have additional reasons for being careful (see Connolly *et al.*, 1991).

Most of the cultural forms with which Britain's black communities have become associated are not long-established traditional cultures brought into Britain by the first settlers then passed down through families. Rastafarianism, dreadlocks, the 'colours', reggae, black art and even samba have been discovered, sometimes invented, by second and third generation British blacks. All peoples need to know who they are, where they are from and why they are in their present conditions. Their answers do not need to be scientifically valid, and their cultures do not need to be long standing in order to be regarded as traditional and to give a people a pride in their heritage and current achievements. So ethnic minority arts, however under-funded in Britain, and blacks' achievements in sport, have enabled them to exhibit and experience pride while surrounded by what they often experience as a hostile society (Dixon, 1991). In a similar way, much of the so-called Scottish heritage is a recent discovery or invention. 'The heritage' has different meanings to different groups of Scots (see McCrone *et al.*, 1995), but this does not diminish its ability to generate and express national pride and identity.

The wider British society's attitudes to ethnic minority cultures have been somewhat ambivalent. On the one hand, there are arguments that British heritage should be prioritized in, for example, state funding for the arts and school syllabuses. Simultaneously, governments have accepted that Britain is now a multi-ethnic and multi-cultural society in which all groups are entitled to equal treatment. Yet liberals worry that support for ethnic minority cultures might contribute to a British version of apartheid. They understand why, but are still uneasy when black and Asian footballers form their own teams and leagues, and when such minority groups demand their own schools. 'Sidetracking' has been another concern; ethnic minorities being channelled into artistic and sporting milieux at the expense of success in mainstream school subjects and employment (Carrington, 1982).

The significance of tourism

None of these processes has been peculiar to Britain and other post-colonial capitalist societies. Nor are any of these processes likely to be eradicated while all peoples, minorities and majorities, use leisure to express national and ethnic identities. Needless to say, the contents of the processes are always distinctive, drawing as they must on the particular histories, cultures and current experiences of the peoples concerned. And all peoples tend to regard their own cultures as special. All this is unchanging. The really distinctive issues surrounding contemporary leisure and national identities arise from the greater volumes and accelerating flows of nationalities and their cultures; much of the current traffic being for leisure purposes.

In the past, for most people, contact with other cultures was extremely limited. Only traders, pilgrims and armies travelled and their journeys took considerable time by modern standards. Air travel, telephones, television and satellite communication have changed all this. They have been responsible for creating the so-called global village and elevating globalization and internationalization into buzz words in present-day social science. There is more, and more rapid, international trade in goods, financial and business services, and in cultural products ranging from clothing fashions through music to languages. People are also travelling further and more frequently. The worldwide drift from less developed to developed countries continues and has been joined by a new flow from East to West Europe. Some of this travel is to settle. Much is temporary economic migration. But there are also considerable flows, in all directions, of international tourists. A result is that virtually all peoples have more contact with others as travellers and residents, and they also have more contact with each other's manufactured goods and cultures even when people themselves remain static.

There is continuing debate about whether an outcome will be a worldwide standardization and homogenization of cultures and lifestyles, and, if so, whether this will be Westernization or Americanization, or whether music, television entertainment and cuisine will become one 'mush' which simultaneously belongs to everybody but nobody in particular. This would finally help to dissolve national and ethnic divisions but at present such an outcome seems unlikely. Up to now the tendency has been for every place and people to produce a distinctive blend of the global and the local.

A benefit of tourism has always been said to be its contribution to international understanding and friendship. Indeed, there is a theory which likens tourism to a modern religion in which we make pilgrimages to 'worship' at each other's 'shrines', return with photographs and other souvenirs to rekindle the memories and prove our devotion, and thereby become members of a common humanity, citizens of the world (MacCannell, 1976). It has always been difficult to reconcile such ideas with the holiday behaviour of the British, and other North Europeans, at Mediterranean resorts. In many ways, tourism gives people incentives to accentuate their national characteristics and heightens their awareness of how they differ from other peoples. However, such is the present economic importance of tourism to many cities, regions and countries that they have good reasons for tolerating and even encouraging visitors' peculiar ways, and presenting their own assets in ways that are attractive to the 'tourist gaze' (Urry, 1990). National and regional heritage, and local arts, come to be valued primarily in so far as they attract visitors. If a place's natural assets are insufficiently attractive, beaches or mountain chalets can be built, historical relics can be discovered, and an inoffensive history can be invented. National majorities have been given an incentive to preserve and present minority cultures in so far as their presence strengthens a place's attractiveness to tourists. Real natives may need to be kept away from the tourist trails lest they blemish the experience. Local citizens' pride in their places and histories may be heightened by the attention of visitors. Alternatively, they may resent being gazed upon or sidelined, and they may feel unable to identify with the heritage that is presented and thereby feel dispossessed of their real identities (Hughes and Boyle, 1992). Modern leisure has added new dimensions to the age-old lifestyle politics of national identities.

International sport

Sport has been given new dimensions by its internationalization. Despite the earlier ideals of the modern Olympic movement, international sport has been structured by national divisions. Success has become important to many people in virtually all countries, and to most governments throughout the world. A result is that players are now adopted by countries almost as easily as they are acquired by clubs named after particular towns and cities.

The crucial point as regards the role of leisure is how easy it is to illustrate that leisure activities are still shaped by wider social divisions and how they express identities with other bases – occupational classes, gender and nationality (see for example, Nauright and Chandler, 1996). There are still no recorded cases, except in postmodern imaginations, of music or sport, for example, becoming bases for comparably enduring and influential identities.

Chapter 9

Leisure: Present and Future

Low Risk Predictions

Making predictions can offer unnecessary hostages to fortune but leisure forecasting is less hazardous than most social science crystal gazing. Leisure behaviour is exceptionally predictable and remains so despite all the talk about flexibilization and the accelerating pace of change which may indeed be destabilizing other parts of people's lives. For many people leisure appears to have become a relatively secure and reliable haven in societies where other things change more rapidly and unpredictably. In modern times, the recent past has been an excellent guide to leisure in the near future. This is because people at leisure tend to be creatures of habit. Chapter 5 noted the extent to which leisure conduct is marked by strong continuities from day to day, week to week, and even from year to year across the life span. Leisure may be less constrained, and may offer more scope for choice than is available in other domains, but most people use their discretion to stick to the familiar.

The main age effects that will modify people's leisure during the life course in the near future are predictable from previous generations. The life events (parenthood, bereavement and so on) that will affect people's leisure over the next 30 years, and what the leisure consequences will be, can be forecast more reliably than whether or when given businesses and occupations will expand or contract. Today people's leisure experiences are less turbulent than their occupational careers, and often their day-to-day working lives. And people do not overhaul their leisure habits as fast as governments and their policies can change.

When new leisure practices spread, this usually happens slowly, first among young people who then carry their new tastes into older age groups. In the recent past this has happened with rock music and soft drug use. Some medium-term developments in the public's uses of leisure are therefore predictable from recent inter-generational trends: the rises in sport participation and computer literacy among young people for example. The medium-term changes in adult leisure that will follow the trends among young people are already 'in the system'.

Other macro-changes usually occur within settled leisure patterns such as when television largely replaced radio listening and going to the cinema.

What all this means is that leisure during the next 20 years is most likely to be much the same as leisure in the 1980s and 1990s, that some trends can be forecast reasonably confidently, and that unexpected changes are unlikely to be lifestyle shattering. Economic growth, if this continues, is certain to lead to a growth of some combination of leisure time, spending and activity, and any growth is most likely to be in the same areas of activity and spending that have expanded in the recent past. It is a fairly safe prediction that continuing economic growth will be accompanied by further rises in spending on holidays and home entertainment.

All this assumes, justifiably if the interpretation of the evidence in previous chapters is correct, that we are not living in an era of wholesale revolutionary upheaval comparable with the birth of modern society when the population shifted from rural to urban areas, from agriculture into industry, and built new ways of life which included modern leisure. There are, of course, important changes in process but these are simply not in the same league as the advent of modernity. For closer similarities to the earlier modernization of the West we have to look towards the developing world countries that are currently modernizing, especially at the people who are moving into modern population centres and jobs, and where tourism is becoming the dominant industry. We can also look towards former communist countries. Globalization has a rather different significance in different places. In some places it really is transforming everyday ways of life. Even the more spectacular leisure trends in the economically advanced societies – trips to Disney World instead of Blackpool Tower for example – are not really of the same order.

Pundits can adopt positions on whether Western civilization is the end of history with little risk of being proved decisively wrong in their own lifetimes. The collapse of communism was accompanied by forecasts of Western capitalism and democracy spreading worldwide, and reigning indefinitely without serious external or internal challenges. The grounds for such triumphalism could prove very short-lived. If the earth and humanity survive for another 1000 years, people may or may not look back upon the 20th century as a highlight of civilization. Whatever their verdict, it is highly likely that the present Western way of life – its forms of paid work, voluntary associations and consumer culture – will have been superseded. No one can be sure. My hunch is that the Western way of life will be of finite duration, but in the closing decades of the 20th century no replacement way of life is yet in sight. Yoga and some Eastern faiths have Western followers. The Prince of Wales regularly indicts Western materialism. He is not alone. There are many in the West who feel that even the fruits of success can be hollow. Nevertheless, this dissent is not yet stretching the economic, social and cultural systems to breaking point. Obtaining the labour and generating the consumption that the development of these systems require are not serious pressure points. Enthusiasm for the Western way of life throughout the rest of the world outweighs the opposing forces.

It is true that the West has problems to which, as yet, there are no known solutions. This applies to the tensions that surround inequalities, unemployment, crime and ecological sustainability. However, the chances are that, in the immediate future as in the recent past, these problems will be tackled without shattering the system. If the problems cannot be solved the system is likely to continue to cope with them.

There is a long, Western, originally Judaic, tradition of forecasting an Armageddon, and during the second half of the 20th century there has been a speed up in the recycling of secular versions of such forecasts. In the 1950s automation was going to decimate paid work and create a seismic explosion in leisure time. In the 1960s post-materialist counter-cultures were going to inspire a flight from paid work and consumption, and an all-round greening of the West. Some believed then that if the West's war machines were wound down the rest of the world would live in peace and friendship. In the 1970s it was the rapid depletion of natural resources and impending ecological disaster that were to puncture the Western dream. Subsequently the return of mass unemployment has revived the belief that an alternative way of life is not just necessary but inevitable. America, as ever, is a land where it is constantly possible to find examples of countercultural movements: against television, materialism in general, and for a return to godliness. However, America's lurch towards Christian fundamentalism has been more concerned with curbing abortion than the business culture. Rather than toppling or contracting, the Western way of life is currently becoming more widespread than ever.

Some current leisure trends in the West are different, but they are not obviously faster or more dramatic than the changes that occurred between 1900 and 1960. After all, during the first half of the 20th century people's ways of life were disrupted by two world wars. The leisure innovations of that era included the radio and the cinema. Some current trends are very long-running: the growth of leisure time, activity and spending, commercialization, the blurring of former social divisions and the related individualization, for instance. During the last 50 years there have been no major leisure developments that have been incomprehensible within existing concepts such as holidays, entertainment and leisure itself, and the chances are that this will remain the case throughout the early decades of the 21st century. Micro-electronics has made a stronger and faster impact in workplaces, often leading to major labour savings, than in most people's leisure lives. Despite all that is written about the Worldwide Web and its revolutionary potential, the impact on leisure seems unlikely to match that of television and the motor car. There have already been Internet courtships leading to marriage but it difficult to envisage email replacing most face-to-face conversation.

Leisure providers may not recognize this tranquillity but there is nothing new in this. Their markets have always been volatile; subject to flights of fashion, and intensely competitive. The providers' core problem, whether they be public or voluntary sector or commercial, is that none can structure and stamp a clear and reliable pattern on leisure demand. It is types of people who structure leisure demand, and the fact that the macro-contours of demand are reasonably stable and predictable is little comfort to suppliers whose viability can be threatened by marginal losses of trade, whose wares are vulnerable to marginal shifts in fashion, and who are constantly threatened by new entrants who can offer something much the same but crucially different in presentation or price. More people taking holidays in the Mediterranean is a threat to traditional holiday camps even though it may not amount to a revolution in leisure habits. The public sector is becoming more exposed to market forces, and voluntary associations have always needed to compete for both members and customers or clients.

Building on the Foundations

The fact that we are not living in an age of lifestyle shattering changes is one reason why leisure research can develop squarely on the foundations laid by past enquiries. Another reason is that past research has been a success story. It is anything but a failed project. It has been successful in charting and explaining how and why leisure has grown, and how its distribution and uses are related to social class, age and gender. Even new activities, such as mass international tourism, have been explicable by building on older theories, in this case about the growth of leisure time and spending. When leisure time and money are redistributed, as is currently happening, explanations can invariably be found by developing existing theories: about the population's time and money preferences, employers' strategies and labour market processes for example.

The theories that leisure research has developed may appear disappointingly weak. But the crucial fact of this matter is that leisure is a field where weak (and flexible) theories yield the strongest explanations. The most effective theories are pluralist. This is so because leisure behaviour is subject to numerous influences related to age, gender, socio-economic status and so on. These influences operate in a multitude of configurations. The end result is that explanations have to vary depending on the leisure activity, the time, the place, and the people who are involved. Stronger theories – certain variants of Marxism and feminism for example – have a simplicity that can generate an instant appeal, but their promise is fraudulent; their explanations prove unsatisfactory. Pluralist theory is also weak on policy implications, but rightly so, because the crucial fact here is that no providers can exert more than a marginal influence on the public's leisure behaviour. Needless to say, a marginal impact across the public in general can be crucial for groups with specific leisure interests whether in land sports, the use of water space or the arts. The weak policy implications do not indicate inadequacies in the theory but simply reflect the reality of modern leisure. There is no need for theoretical upheaval.

That said, there are numerous changes underway which require a non-routine response from leisure research. The changes may not be revolutionary but they are still important to the people whose lives are affected directly. None of the changes is being generated primarily within leisure. Their origins are in the surrounding social and economic contexts, and leisure research is well-equipped to explore the implications for the everyday lives of different sections of the population. The previous chapters have emphasized how the research agendas in all the standard areas need fine-tuning if not more thorough revamps to engage with current issues.

First, we need research in order to reduce the speculation in current explanations of why, in Britain and some other countries, some sections of the workforces are working longer, and the implications for their out-of-work lives. Likewise there is a need for research into the consequences of the destandardization of working time. The effects are unlikely to be exactly the same in all sections of the population. This presumption would fly in the face of everything that we know about age, sex and social class differences. These are issues on which theoretical speculation has raced ahead of our evidence, thereby creating an agenda for a new generation of enquiries and researchers.

Second, we need to learn more about the implications for men's and women's lives of more of the latter entering employment, taking shorter career breaks, working longer hours, and commanding a higher share of total earnings. Even more pressing, there is a dearth of research into the significance for men, and their partners if they have any, of males taking on more housework and the devaluation of traditional brands of masculinity across much of the labour market. Once again, the implications are most likely to differ for both women and men according to their ages and socio-economic status. Upper middle class men and women may be leading more similar lives at work and at home (see Chapter 4). If so, their experiences of macro-trends will be very different from those of males and females in the so-called socially and economically excluded groups wherein high proportions of the women experience lone parent situations and many males find themselves redundant not merely in the labour market but in the domestic sphere also.

Third, we need more systematic exploration and less speculation about the implications of the replacement of predictable life cycle sequences, formerly common among broad classes of males and females, by more individualized life courses: the wider variety within and the increasing overlap in the circumstances of people in different age groups. If individuals' leisure behaviour remains as stable and predictable as argued earlier, how is this achieved when changes in people's work and family lives require changes in their places of residence and in their daily and weekly time schedules, and, therefore, in where their leisure time can be spent and with whom? If people manage to preserve friendships and group memberships, how do they achieve this? We need typologies of individualized biographies that will identify what are likely to be the diverse implications of life course destandardization for men and women in different social classes.

Fourth, the growth of consumption and commercial leisure, and the continuing spread and deepening of consumer culture need to be taken seriously by leisure research. Deconstructing adverts, and making or rejecting extravagant claims about their significance, are insufficient. We need enquiries that situate the impact of commercial provisions and processes within lifestyles that also use public and voluntary sector provisions, and where much leisure is organized privately, and enquiries into how the nuances vary, as they are most likely to do, by age, sex and socio-economic status.

Fifth, the issue of leisure and identity needs to be properly investigated now that we know what questions to ask. The real issue is not whether leisure-based lifestyles are replacing older sources of identity but the significance of short-term and longer-term leisure embellishments and, once again, how these vary between different sections of the population.

Addressing all these issues will require attention to how leisure is experienced. This issue is manageable once researchers realize that they are not required to trawl the whole of human experience but to focus on what people do with their leisure time and money.

Previous research has laid secure and substantial foundations from which all the above lines of enquiry can proceed. Perhaps most important of all, we now know that the most powerful explanations arise through investigating types of people rather than types of leisure. This is why the field of enquiry is impoverished and disempowered theoretically when it splinters into studies of sport, tourism, the media and so on. Society is not neatly divided into sport players, tourists, theatre-goers etc. It is the same people who do all these things. And the most powerful explanations of all their types of leisure behaviour need to take account of the actors' total situations, and the broader ways of life or lifestyles to which specific uses of leisure contribute. It will be in the course of investigating the issues listed above that the significance of the continuing growth of tourism, developments in telecommunications, the spread of personal computers, more people using the Internet, and developments in sport, film, the printed media and so on will be established.

Justifying the Agenda

The reasons given in Chapter 1 for studying leisure are still the best. They have not been superseded by the subsequent evidence and arguments. No new justifications for leisure research are required; just recognition that the old reasons have become even stronger than in the past simply because leisure grows remorselessly in modern societies; not equally so for all sections of the populations but, even so, the unmistakable long-term and continuing trend is towards leisure becoming a larger and weightier part of people's lives.

Voluntary and commercial organizations need to know about their members, clients and customers, actual and potential. Governments need to grasp why they inevitably become embroiled in leisure, the goals that they can realistically pursue and the likely consequences for different sections of the population. Leisure is socially important on account of its capacity to strengthen some groups thereby deepening some social divisions and ameliorating others. Overall the trends in leisure are tending to blur divisions between age groups, the sexes and occupation-based social classes thereby contributing to broader processes of individualization. Leisure continues to perform the recuperative, learning and expressive functions that make it vital for individuals' well-being. Then there is leisure's economic significance as a source of business and employment. Its importance in these respects varies between regions and countries but there are few places today where leisure can be safely ignored in economic policy and planning. These have always been, and remain, important enough reasons for studying leisure.

Bibliography

Abrams, M. (1961) *The Teenage Consumer.* London Press Exchange, London.
Abrams, M. (1977) Quality of life studies. In: Smith, M.A. (ed.) *Leisure in the Urban Society.* Leisure Studies Association, Manchester.
Abrams, M. (1980) Class differences in the use of leisure time by the elderly. *Leisure Studies Association Quarterly* 1(4), 2–3.
Adam, B. (1990) *Time and Social Theory.* Polity Press, Cambridge.
Adomo, T. and Horkheimer, M. (1977) The Culture Industry: enlightenment as mass deception. In: Curran, J., Gurevitch, M. and Woollacott, J. (eds) *Mass Communication and Society.* Edward Arnold, London.
Aitchison, C. (1997) A decade of compulsory competitive tendering in UK sport and leisure services: some feminist reflections. *Leisure Studies* 16, 85–105.
Albermarle Report (1960) *The Youth Service in England and Wales.* HMSO, London.
Allan, G. and Crowe, G. (1991) Privatisation, home centredness and leisure. *Leisure Studies* 10, 19–32.
Andrews, F.M. and Withey, S.B. (1976) *Social Indicators of Well-Being.* Plenum Press, New York.
Ang, I. (1996) *Living Room Wars.* Routledge, London.
Bailey, P. (1978) *Leisure and Class in Victorian England.* Routledge, London.
Banks, M., Bates, I., Breakwell G., Bynner, J., Emler, N., Jamieson, L. and Roberts, K. (1992) *Careers and Identities.* Open University Press, Milton Keynes.
Barrell, G., Chamberlain, A., Evans, J., Holt, T. and Mackean, J. (1989) Ideology and commitment in family life: the case of runners. *Leisure Studies* 8, 249–262.
Baudrillard, J. (1998) *The Consumer Society: Myths and Structures.* Sage, London.
Beatson, M. (1995) *Labour Market Flexibility.* Employment Department Research Series 48, Sheffield.
Beck, U. and Beck-Gernsheim, E. (1995) *The Normal Chaos of Love.* Polity Press, Cambridge.
Bell, C. and Healey, P. (1973) The family and leisure. In: Smith, M.A. *et al.* (eds) *Leisure and Society in Britain.* Allen Lane, London.
Berger, P.A., Steinmuller, P. and Sopp, P. (1993) Differentiation of life courses? Changing patterns of labour market sequences in West Germany. *European Sociological Review* 9, 43–61.
Berrett, T., Burton, T.L. and Slack, T. (1993) Quality products, quality service: factors

leading to entrepreneurial success in the sport and leisure industry. *Leisure Studies* 12, 93–106.

Beute, A.M. and Marini, M.M. (1995) Gender and values. *American Sociological Review* 60, 436–448.

Bianchini, F. and Parkinson, M. (eds) (1993) *Cultural Policy and Urban Regeneration.* Manchester University Press, Manchester.

Bienefeld, M.A. (1972) *Working Hours in British Industry.* Weidenfeld and Nicolson, London.

Bishop, J. and Hoggett, P. (1986) *Organizing Around Enthusiasms.* Comedia, London.

Blaxter, M. (1990) *Health and Lifestyles.* Tavistock/Routledge, London.

Blyton, P. and Trinczek, R. (1996) *The Reincarnation of Worksharing as a Response to Job Cuts: Assessing Recent Developments in Germany.* Hans Bockler Foundation, Dusseldorf.

Bocock, R. (1993) *Consumption.* Routledge, London.

du Bois-Reymond, M., Dickstra, R., Hurrelmann, K. and Peters, E. (eds) (1995) *Childhood and Youth in Germany and The Netherlands.* de Gruyter, Berlin.

Boothby, J., Tungatt, M., Townsend, A.R. and Collins, M.F. (1981) *A Sporting Chance?* Sports Council Study 22, London.

Bosworth, D. (1994) *Sunday Working: An Analysis of an Employer Survey.* Employment Department Research Series 33, Sheffield.

Bourdieu, P. (1984) *Distinction: A Social Critique of the Judgement of Taste.* Routledge, London.

Bourdieu, P. and Darbel, A. (1997) *The Love of Art.* Polity, Oxford.

Bowden, S. (1994) The new consumerism. In: Johnson P. (ed.) *Twentieth Century Britain.* Longman, London.

Bramham, P. and Henry, I.P. (1985) Political ideology and leisure policy in the United kingdom. *Leisure Studies* 4, 1–19.

Breedveld, K. (1994) Post-Fordist leisure and work. Paper presented at International Sociological Association Conference, Bielefeld.

Breedveld, K. (1996a) Working odd hours: revolution in time or storm in a tea-cup? Paper presented at World Leisure and Recreation Association Conference, Cardiff.

Breedveld, K. (1996b) The double myth of flexibilisation: trends in scattered work hours and differences in time sovereignty. Paper presented to the conference on New Strategies for Everyday Life, Tilburg.

Brennan, D. (1993) Adolescent girls and disco dancing. In: Brackenridge, C. (ed.) *Body Matters.* Leisure Studies Association Publication 47, Eastbourne.

Bridges, W. (1995) *Job Shift.* Allen and Unwin, London.

Brook, J. (1993) Leisure meanings and comparisons with work. *Leisure Studies* 12, 149–162.

Brown, D. and Charles, N. (1982) *Women and Shiftwork: Some Evidence from Britain.* European Foundation, Dublin.

Brown, H.G. (1959) Some Effects of Shiftwork on Social and Domestic Life. *Yorkshire Bulletin of Economic and Social Research*, Occasional Paper 2.

Brown, P. (1987) *Schooling Ordinary Kids.* Tavistock, London.

Brown, R.K. (1990) A flexible future in Europe? Changing patterns of employment in the United Kingdom. *British Journal of Sociology* 41, 301–327.

Brown, S. (1995) Crime and safety in whose community? *Youth and Policy* 48, 27–48.

Butler, K.N. (1978) Roles of the commercial provider in leisure. In: Talbot, M.A. and Vickerman R.W. (eds) *Social and Economic Costs and Benefits of Leisure.* Leisure Studies Association, Leeds.

Butler, T. and Savage, M. (eds) (1995) *Social Change and the Middle Classes.* UCL Press, London.

Bynner, J. and Ashford, S. (1992) Teenage careers and leisure lives: an analysis of lifestyles. *Society and Leisure* 15, 499–519.

Cale, L. and Almond, L. (1992) Physical activity levels of secondary aged children: a review. *Health Education Journal* 51, 192–197.

Callois, R. (1955) The structure and classification of games. *Diogenes* 12, 62–75.

Campbell, C. (1995) The sociology of consumption. In: Miller, D. (ed.) *Acknowledging Consumption.* Routledge, London.

Campbell, C. (1997) Shopping, pleasure and the sex war. In: Falk P. and Campbell, C. (eds) *The Shopping Experience.* Sage, London, pp. 167–176.

Carrington, B. (1982) Sport as a sidetrack. In: Barton, L. (ed.) *Class, Race and Gender in Education.* Croom Helm, London.

Carrington, B., Chivers, T. and Williams, T. (1987) Gender, leisure and sport: a case study of young people of South Asian descent. *Leisure Studies* 6, 265–279.

Carter, F.A. and Corlett, E.N. (1982) *Review of the European Foundation's Research into Shiftwork, 1977–80.* European Foundation, Dublin.

Cashmore, E. (1994) *And There was Television.* Routledge, London.

Chambers, D.A. (1983) Symbolic equipment and the objects of leisure images. *Leisure Studies* 2, 301–315.

Champoux, J.E. (1978) Perceptions of work and non-work. *Sociology of Work and Occupations* 5, 402–422.

Chaney, D. (1996) *Lifestyles.* Routledge, London.

Chase, D.R. and Godbey, G.C. (1983) The accuracy of self-reported participation rates. *Leisure Studies* 2, 231–235.

Cherry, G. (1984) Leisure and the home: a review of a changing relationship. *Leisure Studies* 3, 35–52.

Child, E. (1981) Play as a social product. *Leisure Studies Association Quarterly* 2(4), 2–4.

Clarke, J. and Critcher, C. (1985) *The Devil Makes Work.* Macmillan, London.

Cloke, P., Phillips, M. and Thrift, N. (1995) The new middle classes and the constructs of rural living. In: Butler, T. and Savage, M. (eds) *Social Change and the Middle Classes.* UCL Press, London.

Coalter, F. (1998) Leisure studies, leisure policy and social citizenship: the failure of welfare or the limits of welfare? *Leisure Studies* 17, 21–36.

Coalter, F. Long, J. and Duffield, B. (1988) *Recreational Welfare.* Avebury, Aldershot.

Cohen, P. (1976) Subcultural concepts and working class community. In: Hammersley, M. and Woods P. (eds) *The Process of Schooling.* Routledge, London.

Cohen, S. (1972) *Folk Devils and Moral Panics.* MacGibbon and Kee, London.

Coles, B. (1995) *Youth and Social Policy.* UCL Press, London.

Connolly, M., Roberts, K., Ben-Tovim, G. and Torkington, P. (1991) *Black Youth in Liverpool.* Giordano Bruno, Culemborg.

Coppock, V., Haydon, D. and Richter, I. (1995) *The Illusions of Post-Feminism.* Taylor and Francis, London.

Corcoran-Nantes, Y. and Roberts, K. (1995) We've got one of those: the peripheral status of women in male dominated industries. *Gender, Work and Organization*

2, 21–33.
Coveney, I., Jackson, M., Jeffreys, S., Kaye, L. and Mahony, P. (1984) *The Sexuality Papers*. Hutchinson, London.
Cox, B.D. Huppert, F.A. and Whichelow, M.J. (1993) *The Health and Lifestyle Survey: Seven Years On*. Dartmouth, Aldershot.
Crewe, I. (1989) The decline of labour and the decline of Labour: social and electoral trends in post-war Britain. In: *Essex Papers in Government and Politics* vol. 65. University of Essex, Colchester.
Cross, G. (1993) *Time and Money: The Making of Consumer Culture*. Routledge, London.
Csikszentmihalyi, M. (1990) *Flow: The Psychology of Optimal Experience*. Harper and Row, New York.
Csikszentmihalyi, M. (1993) Activity and happiness. *Journal of Occupational Science* 1(1), 38–42.
Cunningham, H., (1980) *Leisure in the Industrial Revolution*. Croom Helm, London.
Cushman, G. Veal, A.J. and Zuzanek, J. (eds) (1996) *World Leisure Participation: Free Time in the Global Village*. CAB International, Wallingford.
Dale, A. (1986) Differences in car usage for married men and married women. *Sociology* 20, 91–92.
Dare, B., Walton, G. and Coe, W. (1987) *Concepts of Leisure in Western Thought*. Kendall/Hunt, Dubuque.
Darton, D. (1986) Leisure forecast 1986: the leisured society. *Leisure Management* January, 7–8.
Davidson, P. (1996) The holiday and work experiences of women with young children. *Leisure Studies* 15, 89–103.
Davies, A. (1992) *Leisure, Gender and Poverty*. Open University Press, Buckingham.
Davies, A. (1994) Cinema and broadcasting. In: Johnson, P. (ed.) *Twentieth Century Britain*. Longman, London.
Dawson, D. (1988a) Leisure and the definition of poverty. *Leisure Studies* 7, 221–231.
Dawson, D. (1988b) Social class in leisure: reproduction and resistance. *Leisure Sciences* 10, 193–202.
Deem, R. (1982) Women, leisure and inequality. *Leisure Studies* 1, 29–46.
Deem, R. (1986) *All Work and No Play?* Open University Press, Milton Keynes.
Deem, R. (1996) Women, the city and holidays. *Leisure Studies* 15, 105–119.
Denzin, N.K. (1991) *Images of Postmodernism*. Sage, London.
Department for Education (1995) *Young People's Participation in the Youth Service*. Statistical Bulletin 1/95, London.
Devine, F. (1992) *Affluent Workers Revisited? Privatism and the Working Class*. Edinburgh University Press, Edinburgh.
Dixon, R.M. (1991) *Black Arts, Policy and the Issue of Equity*. Race and Social Policy Unit, University of Liverpool.
Dumazedier, J. (1967) *Towards a Society of Leisure*. Free Press, New York.
Dumazedier, J. (1974) *Sociology of Leisure*. Elsevier, Amsterdam.
Dumazedier, J. (1989) France: leisure sociology in the 1980s. In: Olszewska, A. and Roberts, K. (eds) *Leisure and Lifestyle*. Sage, London.
Duncombe, J. and Marsden, D. (1993) Love and intimacy: the gender division of emotion and emotion work. *Sociology* 27, 221–241.
Dunning, E. (1996) On problems of the emotions in sport and leisure: critical and counter-critical comments on the conventional and figurational sociologies of

sport and leisure. *Leisure Studies* 15, 185–207.

Dunning, E. and Rojek, C. (eds) (1992) *Sport and Leisure in the Civilising Process.* Macmillan, Basingstoke.

Dunning, E. Murphy, P. and Williams, J. (1986) Spectator violence at football matches: towards a sociological explanation. *British Journal of Sociology.* 37, 221–244.

Dunning, E., Murphy, P. and Waddington, L. (1992) *Violence in the British Civilising Process.* Discussion Papers in Sociology S92/2, University of Leicester.

Eichenbaum, L. and Orbach, S. (1984) *What Do Women Want?* Fontana, London.

van Eijek, K. and van Rees, K. (1998) The impact of social mobility on patterns of cultural consumption: individual omnivores and heterogeneous status groups. Paper presented to International Sociological Association Congress, Montreal.

Emler, N. and McNamara, S. (1996) The social contact patterns of young people: effects of participation in the social institutions of family, education and work. In: Helve, H. and Bynner, J. (eds) *Youth and Life Management: Research Perspectives.* Helsinki University Press, Yliopistopaino.

Erickson, B.H. (1996) Culture, class and connections. *American Journal of Sociology* 102, 217–251.

Estes, R.J. and Wilenski, H. (1978) Life-cycle squeeze and the morale curve. *Institute of Industrial Relations, Reprint 422.* University of California, Berkeley.

European Foundation for the Improvement of Living and Working Conditions (1980) *The Effects of Shiftwork on Health, Social and Family Life.* European Foundation, Dublin.

Evans, S.T. and Haworth, J.T. (1991) Variations in personal activity, access to categories of experience, and psychological well-being in young adults. *Leisure Studies* 10, 249–264.

Eygendaal, W. (1992) The black heart – a qualitative study of the death metal culture in The Netherlands. Paper presented to the Conference on Internationalisation and Leisure Research, Tilburg.

Fache, W. (1996) The common weekend is threatened in Belgium. Paper presented to the Conference on New Strategies for Everyday Life, Tilburg.

Fajertag, G. (1996) Working time policies in Europe: recent trends. Paper presented to the Conference on New Strategies for Everyday Life, Tilburg.

Featherstone, M. (ed.) (1988) *Postmodernism: Theory, Culture and Society*, Vol. 5, Nos 2–3. Sage, London.

Featherstone, M. (1991) *Consumer Culture and Post-Modernism.* Sage, London.

Fishwick, L. and Hayes, D. (1989) Sport for whom? Differential participation patterns of recreational athletes in leisure time physical activities. *Sociology of Sport Journal* 6, 269–277.

Fleming, S. (1995) *Home and Away: Sport and South Asian Male Youth.* Avebury, Aldershot.

Freysinger, V.J. and Chen, T. (1993) Family and leisure in China: the impact of culture. *World Leisure and Recreation* 35(3), 22–24.

Fryer, D. and Payne, R. (1984) Proactive behaviour in unemployment: findings and implications. *Leisure Studies* 3, 273–295.

Furlong, A., Campbell, R. and Roberts, K. (1990) The effects of post-16 experiences and social class on the leisure patterns of young adults. *Leisure Studies* 9, 213–224.

Gabriel, Y. and Lang, T. (1995) *The Unmanageable Consumer: Contemporary Consumption and its Fragmentation.* Sage, London.

Gallie, D., Marsh, C. and Vogler, C. (eds) (1994) *Social Change and the Experience of Unemployment.* Oxford University Press, Oxford.

Gattas, J.T., Roberts, K., Schmitz-Scherzer, R., Tokarski, W. and Vitanyi, I. (1987) Leisure and lifestyles: towards a research agenda. *Society and Leisure* 9, 529–539.

Gershuny, J.J. (1986) Leisure: feast or famine? *Society and Leisure* 9, 431–454.

Gershuny, J. (1992) Are we running out of time? *Futures* January/February, 3–22.

Gerth, H.H. and Wright Mills, C. (1948) *From Max Weber.* Routledge, London.

Giddens, A. (1991) *Modernity and Self-Identity.* Polity Press, Cambridge.

Ginn, J., Arber, S., Brannen, J., Dale, A., Dex, S., Elias, P., Moss, P., Pahl, J., Roberts, C. and Rubery, J. (1996) Feminist fallacies: a reply to Hakim on women's employment. *British Journal of Sociology* 47, 167–174.

Glyptis, S. (1981) Leisure life-styles. *Regional Studies* 15, 311–326.

Glyptis, S. (1989) *Leisure and Unemployment.* Open University Press, Milton Keynes.

Glyptis, S.A. and Chambers, D.A. (1982) No place like home? *Leisure Studies* 1, 247–262.

Glyptis, S., McInnes, H. and Patmore, J.A. (1987) *Leisure and the Home.* Sports Council/Economic and Social Research Council, London.

Godbey, G. (1975) Anti-leisure and public recreation policy. In: Parker, S. *et al.* (eds) *Sport and Leisure in Contemporary Society.* Leisure Studies Association, London.

Goddard, E. (1991) *Drinking in England and Wales in the Late-1980s.* HMSO, London.

Goldthorpe, J.H. (1996) Class analysis and the reorientation of class theory: the case of persisting differentials in educational attainment. *British Journal of Sociology* 47, 481–505.

Goldthorpe, J.H., Lockwood, D., Bechhofer, E. and Platt, J. (1969) *The Affluent Worker in the Class Structure.* Cambridge University Press, London.

Goodale, T. and Godbey, G. (1988) *The Evolution of Leisure.* Venture, State College, Pennsylvania.

Gratton, C. (1992) A perspective on European leisure markets. Paper presented to the Conference on Internationalisation and Leisure Research, Tilburg.

de Grazia, S. (1962) *Of Time, Work and Leisure.* Twentieth Century Fund, New York.

de Grazia, V. (1992) Leisure and citizenship. In: *Leisure and New Citizenship.* Actas VIII Congreso ELRA, Bilbao.

Greater London Council (1986) *A Sporting Chance.* GLC, London.

Green, E. (1998) Women doing friendship: an analysis of women's leisure as a site of identity construction, empowerment and resistance. *Leisure Studies* 17, 171–185.

Green, E., Hebron, S. and Woodward, D. (1990) *Women's Leisure, What Leisure?* Macmillan, London.

Gregory, S. (1982) Women among others: another view. *Leisure Studies* 1, 47–52.

Griffin, C. (1985) *Typical Girls?* Routledge, London.

Griffiths, V. (1995) *Adolescent Girls and their Friends.* Avebury, Aldershot.

Gvozdeva, G.P. (1994) Changes in free time utilization by rural residents in West Siberia under the ongoing economic reform. Paper presented to the International Sociological Association Conference Bielefeld.

Hakim, C. (1993) The myth of rising female employment. *Work, Employment and Society* 7, 97–120.

Hakim, C. (1995) Fine feminist fallacies about women's employment. *British Journal of Sociology* 46, 429–455.

Halford, S. and Savage, M. (1995) Restructuring organizations, changing people: gender and restructuring in banking and local government. *Work, Employment and Society* 9, 97–122.

Hall, J. and Perry, N. (1974) *Aspects of Leisure in Two Industrial Cities*, Occasional Papers in Survey Research, 5. Social Science Research Council, London.

Hall, S. and Jefferson, T. (eds) (1976) *Resistance Through Rituals.* Hutchinson, London.

Hamilton-Smith, E. (1992) Work, leisure and optimal experience. *Leisure Studies* 11, 243–256.

Hantrais, L. (1985) Leisure lifestyles and the synchronisation of family schedules: a Franco-British comparative perspective. *World Leisure and Recreation* 20(2), 18–24.

Hantrais, L. and Kamphorst, T.J. (1987) *Trends in the Arts: A Multinational Perspective.* Giordano Bruno, Amersfoort.

Harada, M. (1994) Towards a renaissance of leisure in Japan. *Leisure Studies* 13, 277–287.

Hardey, M. (1990) Family form and leisure opportunities. In: Long, J. (ed.) *Leisure, Health and Well-Being.* Leisure Studies Association, Brighton.

Hargreaves, Jennifer (1982) *Sport, Culture and Ideology.* Routledge, London.

Hargreaves, Jennifer (1994) *Sporting Females.* Routledge, London.

Hargreaves, John (1986) *Sport, Power and Culture.* Polity Press, Cambridge.

Harper, W. (1997) The future of leisure: making leisure work. *Leisure Studies* 16, 189–198.

Havighurst, R.J. and Feigenbaum, K. (1959) Leisure and lifestyle. *American Journal of Sociology* 64, 396–405.

Haworth, J.T. (1993) Skill challenge relationships and psychological well-being in everyday life. *Society and Leisure* 16, 115–128.

Haworth, J.T. and Drucker, J. (1991) Psychological wellbeing and access to categories of experience in unemployed young adults. *Leisure Studies* 10, 265–274.

Hebdige, D. (1979) *Sub-Culture: the Meaning of Style.* Methuen, London.

Hedges, B. (1986) *Personal Leisure Histories.* Economic and Social Research Council/Sports Council, London.

Helve, H. (1998) Attitudes and values of young people and cultural, economic and political change. Paper presented to the International Sociological Association Congress, Montreal.

Hemingway, J.I. (1988) Leisure and civility: reflections on a Greek ideal. *Leisure Sciences* 10, 179–191.

Henderson, K., Bialeschki, M.D., Shaw, S.C. and Freysinger, V.J. (1989) *A Leisure of One's Own.* Venture, Pennsylvania.

Hendry, L.B., Raymond, M. and Stewart, C. (1984) Unemployment, school and leisure: an adolescent study. *Leisure Studies* 3, 175–187.

Hendry, L.B., Shucksmith, J., Love, J.G. and Glendinning, A. (1993) *Young People's Leisure and Lifestyles.* Routledge, London.

Henley Centre (1993) *Inbound Tourism – A Packaged Future.* Henley Centre, London.

Henry, I.P. (1993) *The Politics of Leisure Policy.* Macmillan, Basingstoke.

Hewitt, P.B. (1993) *About Time: The Revolution in Work and Family Life.* IPPR/Rivers

Oram Press, London.
Hey, V. (1986) *Patriarchy and Pub Culture*. Tavistock, London.
Hidy, P. (1982) *Who are being Entertained?* Institute for Culture, Budapest.
Hillman, M. (1991) *One False Move*. Policy Studies Institute, London.
Hinrichs, K., Roche, W. and Sirianni, C. (1991) *Working Time in Transition*. Temple University Press, Philadelphia.
Hirsch, F. (1977) *The Social Limits to Growth*. Routledge, London.
Hobcraft, J. and Kieman, K. (1995) *Becoming a Parent in Europe*. Welfare State Programme 116, London School of Economics.
Hobson, D. (1979) Working class women, the family and leisure. In: Strelitz, Z. (ed.) *Leisure and Family Diversity*. Leisure Studies Association Conference Papers 9, London.
Hodgson, P. (1988) Why leisure research is different. 41st ESOMAR Market Research Conference, Lisbon.
Hoffman, R. (1996) On the road to lifetime working hours. Paper presented to the Conference on New Strategies for Everyday Life, Tilburg.
Hoggart, R. (1957) *The Uses of Literacy*. Chatto and Windus, London.
Hollands, R.G. (1995) *Friday Night, Saturday Night*. Department of Social Policy, University of Newcastle.
Holliday, S. (1996) Trends in British work, leisure and the quality of life. Paper presented to the Conference on New Strategies for Everyday Life, Tilburg.
Horning, K.H., Gerhard, A. and Michailow, M. (1995) *Time Pioneers: Flexible Working Time and New Lifestyles*. Polity Press, Cambridge.
Howe, C.Z. and Rancourt, A.M. (1990) The importance of definitions of selected concepts for leisure enquiry. *Leisure Sciences* 12, 395–406.
Howkins, A. and Lowerson, J. (1979) *Trends in Leisure, 1919–1939*. Social Science Research Council/Sports Council, London.
Hughes, G. (1993) The self, signification and the superyacht. *Leisure Studies* 12, 253–265.
Hughes, G. and Boyle, M. (1992) Place boosterism: political contention, leisure and culture in Glasgow. In: Sugden, J. and Knox, C. (eds) *Leisure in the 1990s: Rolling Back the Welfare State*. Leisure Studies Association Publication 46, Eastbourne.
Huizinga, J. (1949) *Homo Ludens*. Routledge, London.
Hultsman, J. (1995) Spelling leisure. *Leisure Studies* 14, 87–101.
Hunnicutt, B.K. (1988) *Work Without End*. Temple University Press, Philadelphia.
Hutton, W. (1995) High risk. *The Guardian* 30 October.
Huws, U. (1993) *Teleworking in Britain*. Employment Department Research Series 18, Sheffield.
Ibrahim, H. (1982) Leisure and Islam. *Leisure Studies* 1, 197–210.
Ingham, A.G. (1985) From public issue to personal trouble: well-being and the fiscal crisis of the state. *Sociology of Sport Journal* 2, 43–55.
Inglehart, R. (1997) *Modernization and Postmodernization: Cultural, Economic and Political Change in 43 Societies*. Princeton University Press, New Jersey.
Inkson, K. and Coe, T. (1993) *Are Career Ladders Disappearing?* Institute of Management, London.
Institute of Management (1993) *Managers Under Stress*. Institute of Management, London.
Irwin, S. (1995) *Rights of Passage*. UCL Press, London.
Jackson, P.R. and Taylor, P.E. (1994) Factors associated with employment status in

later life. *Work, Employment and Society* 8, 553–567.

Jahoda, M. (1982) *Employment and Unemployment: A Social-Psychological Analysis.* Cambridge University Press, Cambridge.

Jansen-Verbeke, M. (1987) Women, shopping and leisure. *Leisure Studies* 6, 71–86.

Jenkins, C. and Sherman, B. (1981) *The Leisure Shock.* Methuen, London.

Jones, G. (1995) *Leaving Home.* Open University Press, Buckingham.

Jones, S. (1986) *Workers at Play.* Routledge, London.

Jung, B. (1990) The impact of the crisis on leisure patterns in Poland. *Leisure Studies* 9, 95–105.

Jung, B. (1994) For what leisure? The role of culture and recreation in post-communist Poland. *Leisure Studies* 13, 1–15.

Kaplan, M. (1979) *Leisure: Lifestyle and Lifespan.* WB Saunders, Philadelphia.

Karsten, L. (1995) Women's leisure: divergence, reconceptualisation and change. *Leisure Studies* 14, 186–201.

Kay, T.A. (1987) Leisure in the lifestyles of unemployed people: a case study in Leicester. PhD thesis, University of Loughborough.

Kay, T. (1996) Women's work and women's worth: the leisure implications of women's changing employment patterns. *Leisure Studies* 15, 49–64.

Kelly, J. (1983) *Leisure Identities and Interactions.* Allen and Unwin, London.

Kelly, J.R. (1986) Commodification of leisure: trend or tract? *Society and Leisure* 9, 455–475.

Kelly, J.R. (1987) *Freedom To Be: A New Sociology of Leisure.* Macmillan, New York.

Kelly, J.R. (1991) Commodification and consciousness: an initial study. *Leisure Studies* 10, 7–18.

Kelly, J.R. (1994) The symbolic interaction metaphor and leisure: critical challenges. *Leisure Studies* 13, 81–96.

Kelly, J.R. and Pesavento-Raymond, L.C. (1988) *Leisure Activities of Unemployed Black and Hispanic Urban Youth.* University of Illinois at Urbana-Champaign.

Kelly, J.R., Steinkamp, M.W. and Kelly, J.R. (1987) Later-life satisfaction: does leisure contribute? *Leisure Sciences* 9, 189–200.

Kelvin, P., Dewberry, C. and Morley-Bunker, N. (1984) *Unemployment and Leisure.* University College, London.

Kiernan, K.E. (1995) *Transition to Parenthood: Young Mothers, Young Fathers – Associated Factors and Later Life Experiences.* Welfare State Programme 113, London School of Economics.

Kilpatrick, R. and Trew, K. (1985) Lifestyles and well-being among unemployed men in Northern Ireland. *Journal of Occupational Psychology* 58, 207–216.

Kleiber, D.A., Caldwell, L.L. and Shaw, S.M. (1993) Leisure meanings in adolescence. *Society and Leisure* 16, 99–114.

te Kloetze, J.W. (1998) Between freedom and commitment: the post-modern family discovered. Paper presented to the International Sociological Association Congress Montreal.

Knox, C. (1987) Territorialism, leisure and community centres in Northern Ireland. *Leisure Studies* 6, 251–264.

Knulst, M. (1992) An elitist rearguard. *The Netherlands Journal of Social Science* 72–94.

Koch-Weser, E. (1990) A framework for the quantitative study of leisure styles. Paper presented to the International Sociological Association Conference, Madrid.

Kohli, M., Rein, M., Guillemard, A-M. and van Gunsteren, H. (eds) (1992) *Time for Retirement.* Cambridge University Press, Cambridge.

ker, S. (1981) Change, flexibility, spontaneity and self-determination in leisure. Social Forces 60, 323–331.
ker, S. (1983) *Leisure and Work*. Allen and Unwin, London.
ker, S., Hamilton-Smith, E. and Davidson, P. (1993) Serious and other leisure: thirty Australians. *World Leisure and Recreation* 35(1), 14–18.
y, N.C.A. and Johnson, D. (1974) *Leisure and Social Structure*. Hatfield Polytechnic.
son, L.F. (1977) *Working Life and Leisure*. Sunderland Polytechnic.
ka, P. (1996) Experimenting 6 + 6 shift work in Finland. Paper presented to the Conference on New Strategies for Everyday Life, Tilburg.
R., Rose, M. and Rubery, J. (eds) (1994) *Skill and Occupational Change*. xford University Press, Oxford.
n, R.A. and Kern, P.M. (1996) Changing highbrow taste: from snob to omni-re. *American Sociological Review* 61, 900–907.
D.J. (1967) Social participation and happiness. *American Journal of iology* 72, 479–488.
k, W. (1991) Distinctions of fun, enjoyment and leisure. *Leisure Studies* 10, –144.
oel, H. (1994) The modularisation of daily life. In: Henry, I. (ed.) *Leisure, ernity, Postmodernity and Lifestyles*. Leisure Studies Association, urne.
ed.) (1991) *Farewell to Flexibility*. Blackwell, Oxford.
(1996) Moral panics revisited. In: Helve, H. and Bynner, J. (eds) *Youth fe Management: Research Perspectives*. Helsinki University Press, naino.
2) *4 Days, 40 Hours*. Pan, London.
. (1988) The social meanings of leisure. *International Sociology* 3,
and Rapoport, R.N. (1975) *Leisure and the Family Life-Cycle*. London.
Phoenix, A. (1997) Rethinking youth identities: modernist and post-rameworks. In: Bynner, J., Chisholm, L. and Furlong, A. (eds) *Youth, and Social Change in a European Context*. Ashgate, Aldershot.
1998) The changing regulation of public leisure provision. *Leisure* 38–154.
84) The effects of unemployment on the leisure activity participa-ployed steelworkers. In: *Le Temps Libre et le Loisir*. Actes du ial de Recherche, Marly-le Roi.
oviet Sport. Blackwell, Oxford.
isure: the state and the individual in the USSR. *Leisure Studies*
m communist forum to capitalist market – East European sport opean Physical Education Review 1, 15–26.
cDonaldization of Society. Pine Forge Press, Thousand Oaks,
cDonaldization Thesis. Sage, London.
g leisure. *Leisure Studies* 6, 87–91.
re and sociological theory in Britain. *Society and Leisure* 13,
g people, schools, sport and government policies. *Sport,*

Koseki, S. (1989) Japan: homo ludens Japonicus. In: Olszewska, A. and Roberts, K. (eds) *Leisure and Lifestyle*. Sage, London.
Laczko, F. and Phillipson, C. (1992) *Changing Work and Retirement*. Open University Press, Milton Keynes.
Laermans, R. (1994) Leisure as making time: some sociological reflections on the paradoxical outcomes of individualisation. In: Actas do Congreso Mundial do Lazer, *New Routes for Leisure*. Instituto de Ciencias Socias, University of Lisbon.
Lane, R.F. (1991) *The Market Experience*. Cambridge University Press, Cambridge.
Lash, S. (1990) *Sociology of Postmodernism*. Routledge, London.
Lash, S. and Urry, J. (1994) *Economies of Signs and Space*. Sage, London.
Leaman, O. (1984) *Sit on the Sidelines and Watch the Boys Play*. Schools Council Programme Pamphlet, Longman Resources Trust, York.
Lees, S. (1986) *Losing Out*. Hutchinson, London.
Lenskyj, H. (1988) Measured time, women, sport and leisure. *Leisure Studies* 7, 233–240.
Leonard, D. (1980) *Sex and Generation*. Tavistock, London.
Lewis, J.D. and Weigert, A.J. (1981) The structures and meanings of social time. *Social Forces* 60, 432–462.
Linder, S. (1970) *The Harried Leisure Class*. Columbia University Press, New York.
van der Lippe, T. (1996) Trends in time use of men and women. Paper presented to the Conference on New Strategies for Everyday Life, Tilburg.
Lloyd, N. (ed.) (1986) *Work and Leisure in the 1980s*. Sports Council/Economic and Social Research Council, London.
Lobo, F. (1993) Late career unemployment, leisure and lifestyle. Paper presented to the World Leisure and Recreation Association Conference, Jaipur.
Long, J. and Wimbush, E. (1985) *Continuity and Change: Leisure Around Retirement*. Economic and Social Research Council/Sports Council, London.
Longhurst, B. (1996) *Popular Music and Society*. Polity Press, Cambridge.
Lovell, T. (1991) Sport, racism and young women. In: Jarvie, G. (ed.) *Sport, Racism and Ethnicity*. Falmer, London.
Lunt, P.K. and Livingstone, S.M. (1992) *Mass Consumption and Personal Identity*. Open University Press, Milton Keynes.
Lury, C. (1996) *Consumer Culture*. Rutgers University Press, New Brunswick.
Lynch, R. and Veal, A.J. (1996) *Australian Leisure*. Longman, Melbourne.
MacCannell, D. (1976) *The Tourist: A New Theory of the Leisure Class*. Macmillan, London.
McCrone, D., Morris, A. and Keily, R. (1995) *Scotland – the Brand: The Making of Scottish Heritage*. Edinburgh University Press, Edinburgh.
McGlone, A.M. and Pudney, S.E. (1986) Personal consumption, gender and marital status. *Sociology* 20, 88–90.
McGoldrick, A.E. (1983) Company early retirement schemes and private pensions schemes: scope for leisure and new lifestyles. *Leisure Studies* 2, 187–202.
McGuire, F.A., Dottavio, F.D. and O'Leary, J.T. (1987) The relationship of early life experiences to later life leisure involvement. *Leisure Sciences* 9, 251–257.
McKeever, E. (1993) Eating out – a family affair. *World Leisure and Recreation* 35(3), 35–38.
Mac an Ghail, M. (1996) What about the boys? Schooling, class and crisis masculinity. *Sociological Review* 44, 381–397.
Madigan, R. and Munro, M. (1996) House beautiful: style and consumption in the

home. *Sociology* 30, 41–57.
Maffesoli, M. (1994) *The Time of the Tribes*. Sage, London.
Maguire, J. (1991) Sport, racism and British society: a sociological study of elite male Afro/Caribbean soccer and rugby union players. In: Jarvie, G. (ed.) *Sport, Racism and Ethnicity*. Falmer, London.
Malcolmson, R.W. (1973) *Popular Recreations in English Society, 1700–1850*. Cambridge University Press, London.
Mander, J. (1980) *Four Arguments for the Elimination of Television*. Harvester Press, Brighton.
Marsden, D. (1982) *Workless*. Croom Helm, London.
Marsh, A. (1979) *Women and Shiftwork*. HMSO, London.
Marshall, G., Rose, D., Newby, H. and Vogler, C. (1988) *Social Class in Modern Britain*. Hutchinson, London.
Martin, B. and Mason, S. (1986) Spending patterns show new leisure priorities. *Leisure Studies* 5, 233–236.
Martin, B. and Mason, S. (1990) Leisure in a less buoyant economy. *Leisure Studies* 9, 1–6.
Martin, B. and Mason, S. (1992) Current trends in leisure: the changing face of leisure provision. *Leisure Studies* 11, 81–86.
Martin, W.B. and Mason, S. (1998) *Transforming the Futue: Rethinking Free Time and Work*. Leisure Consultants, Sudbury.
Mason, J. (1988) No peace for the wicked: older married women and leisure. In: Wimbush E. and Talbot, M. (eds) *Relative Freedoms*. Open University Press, Milton Keynes.
Mason, T. (1994) Sport and recreation. In: Johnson, P. (ed.) *Twentieth Century Britain*. Longman, London.
Measham, F., Newcombe, R. and Parker, H. (1994) The normalisation of recreational drug use among young people in North-West England. *British Journal of Sociology* 45, 287–312.
Melendez, N. (1992) Life satisfaction and leisure activity patterns among the retired elderly in Puerto Rico. Paper presented to the International Sociological Association Conference, New Routes for Leisure, Lisbon.
Meller, H.E. (1976) *Leisure and the Changing City 1870–1914*. Routledge, London.
Michaelson, M. (1979) The moral dimension of leisure. In: Strelitz, Z. (ed.) *Leisure and Family Diversity*. Leisure Studies Association Conference Papers 9, London.
Mihalik, B.J., O'Leary, J.T., Maguire, F.A. and Dottavio, T.D. (1989) Sports involvement across the life span: expansion and contraction of sports activities. *Research Quarterly for Exercise and Sport* 60, 396–398.
Miles, S., Cliff, D. and Burr, V. (1998) Fitting in and sticking out: consumption, consumer meanings and the construction of young people's identities. *Journal of Youth Studies* 1, 81–96.
Miller, D. (1995) Consumption as the vanguard of history. In: Miller, D. (ed.) *Acknowledging Consumption*. Routledge, London.
Miller, K.A. and Kohn, M.I. (1983) The reciprocal effects of job conditions and the intellectuality of leisure time activities. In: Parker, S. (ed.) *Leisure, Work and Family*. Leisure Studies Association, London.
Mobily, K.E. (1989) Meanings of leisure and recreation among adolescents. *Leisure Studies* 8, 11–23.
Mogenson, G.V. (1990) *Time and Consumption*. Danmarks Statistik, Copenhagen.
van Moorst, H. (1982) Leisure and social theory. *Leisure Studies* 1, 157–169.

Morgan, D.J. (1992) *Discovering Men*. Routledge, London.
Morley, D. (1986) *Family Television: Cultural Power and* Comedia, London.
Mort, F. (1996) *Cultures of Consumption: Masculinities and S Twentieth Century Britain*. Routledge, London.
Mott, P.E. *et al.* (1965) *Shift Work: The Social, Psycho Consequences*. University of Michigan Press.
Mott, J. (1973) Miners, weavers and pigeon racing. In: *Leisure and Society in Britain*. Allen Lane, London.
Mouzelis, N. (1995) *Sociological Theory: What Went Wr*
Mulgan, G. and Wilkinson, H. (1995) Well-being and 2–11.
Mulkay, M. (1988) *On Humour: Its Nature and Plac* Press, Oxford.
Mulkay, M. and Howe, G. (1994) Laughter for s 481–500.
Mullett, S. (1988) Leisure and consumption: incom 7, 241–253.
Mungham, G. and Pearson, G. (eds) (1976) *Work* London.
Murdock, G. (1994) New times/hard times: le good. *Leisure Studies* 13, 239–248.
Myerscough, J. (1974) The recent history of (ed.) *Leisure Research and Policy*. Scot
Nardi, P.M. (ed.) (1992) *Men's Friendships*
Nare, S. (1996) Girls' and boys' econom gender system. In: Helve, H. and B *Research Perspectives*. Helsinki Ur
Nauright, J. and Chandler, T.J.L. (1996) Frank Cass, London.
Netz, Y. (1996) Time allocation an Samuel, N. (ed.) *Women, Lei* CAB International, Wallingfor
Neulinger, J. (1990) *Eden After Al*
Nickson, D., Warhurst, C. Witz, service economy: an ov International Labour Mark
Noon, M. and Blyton, P. (199
O'Connor, B. and Boyle, R. (male and female pleas
Oliver, J. (1998) Losing co
van Ophem, J. and de Ho and the rich in the J.W. (ed.) *Family a* Apeldoorn.
Pahl, J. (1990) Househ marriage. *Socio*
Pahl, R. (1995) *Afte*
Parker, S. (1971) *T*
Parker, S. (1979) R

Education and Society 1, 47–57.

Roberts, K. and Brodie, D. (1992) *Inner-City Sport: Who Plays and What are the Benefits?* Giordano Bruno, Culemborg.

Roberts, K. and Chambers, D.A. (1985) Changing times: hours of work/patterns of leisure. *World Leisure and Recreation* 27(1), 17–23.

Roberts, K. and Jung, B. (1995) *Poland's First Post-Communist Generation.* Avebury, Aldershot.

Roberts, K. and Parsell, G. (1991) Young people's sources and levels of income, and patterns of consumption in Britain in the late-1980s. *Youth and Policy* 35, 20–25.

Roberts, K. and Parsell, G. (1992) Entering the labour market in Britain: the survival of traditional opportunity structures. *Sociological Review* 30, 727–753.

Roberts, K. and Parsell, G. (1994) Youth cultures in Britain: the middle class take-over. *Leisure Studies* 13, 33–48.

Roberts, K., Noble, M. and Duggan, J. (1982) Youth unemployment: an old problem or a new lifestyle? *Leisure Studies* 1, 171–182.

Roberts, K. Brodie, D. and Dench, S. (1987) Youth unemployment and out-of-home recreation. *Society and Leisure* 10, 281–294.

Roberts, K. York, C. and Brodie, D.A. (1988) Participant sport in the commercial sector. *Leisure Studies* 7, 145–157.

Roberts, K., Dench, S., Minten, J. and York, C. (1989a) *Community Response to Leisure Centre Provision in Belfast.* Sports Council Study 34, London.

Roberts, K. Lamb, K.L., Dench, S. and Brodie, D.A. (1989b) Leisure patterns, health status and employment status. *Leisure Studies* 8, 229–235.

Roberts, K. Campbell, R. and Furlong, A. (1990) Class and gender divisions among young adults at leisure. In: Wallace, C. and Cross M. (eds) *Youth in Transition.* Falmer, London.

Roberts, K., Minten, J.H., Chadwick, C., Lamb, K.L. and Brodie, D.A. (1991a) Sporting lives: a case study of leisure careers. *Society and Leisure* 14, 261–284.

Roberts, K. Parsell, G. and Chadwick, C. (1991b) Unemployment and young people's leisure in Liverpool and Swindon. *Society and Leisure* 14, 513–530.

Robinson, J.P. and Godbey, G. (1996) Time inequalities and irrelevancies. Paper presented to the Conference on New Strategies for Everyday Life, Tilburg.

Rojek, C. (1984) Did Marx have a theory of leisure? *Leisure Studies* 3, 163–174.

Rojek, C. (1985) *Capitalism and Leisure Theory.* Tavistock, London.

Rojek, C. (1993) *Ways of Escape.* Macmillan, London.

Rojek, C. (1995) *Decentring Leisure.* Sage, London.

Rojek, C. (1997) Leisure in the writings of Walter Benjamin. *Leisure Studies* 16, 155–171.

Rojek, C. and Urry, J. (eds) (1997) *Touring Cultures: Transformations of Travel and Theory.* Routledge, London.

Rosducher, J. and Seifert, H. (1996) The reduction of working hours and employment: reduction of working hours in Germany and its significance for employment policy. Hans-Bockler Foundation, Dusseldorf.

Rosenweig, R. (1983) *Eight Hours For What We Will.* Cambridge University Press, New York.

Rowe, D. (1995) *Popular Cultures: Rock Music, Sport and the Politics of Pleasure.* Sage, London.

Salaman, G. (1974) *Community and Occupation.* Cambridge University Press, London.

Samdahl, D.M. (1988) A symbolic interactionist model of leisure: theory and empirical support. *Leisure Sciences* 10, 27–39.

Samuel, N. (1996) Introduction. In: Samuel, N. (ed.) *Women, Leisure and the Family in Contemporary Society*. CAB International, Wallingford.

Saunders, P. (1990) *A Nation of Home Owners*. Unwin Hyman, London.

Savage, M. Barlow, J. Dickens, P. and Fielding, T. (1992) *Property, Bureaucracy and Culture*. Routledge, London.

Schor, J. (1998) Beyond work and spend. *Vrijetijd Studies* 18, 7–20.

Schor, J.B. (1991) *The Overworked American*. Basic Books, New York.

Scott, D. and Willits, F.K. (1989) Adolescent and adult leisure patterns: a 37 year follow-up study. *Leisure Sciences* 11, 323–335.

Scraton, S. (1987) Boys muscle in where angels fear to tread – girls' subcultures and physical activities. In: Horne, J. Jary, D. and Tomlinson, A. (eds) *Sport, Leisure and Social Relations*. Routledge, London.

Scraton, S. (1992) *Shaping up to Womanhood: Gender and Girls' Physical Education*. Open University Press, Buckingham.

Scraton, S. (1994) The changing world of women and leisure: feminism, post-feminism and leisure. *Leisure Studies* 13, 249–261.

Seabrook, J. (1988) *The Leisure Society*. Blackwell, Oxford.

Shamir, B. (1985) Unemployment and free time–the role of the protestant work ethic and work involvement. *Leisure Studies* 4, 333–345.

Sharkey, A. (1997) The land of the free. *Weekend Guardian* 22 November, 14–25.

Sharp, D.J. Greer, J.M. and Lowe, G. (1988) The normalization of under-age drinking. Paper presented to the British Psychological Society, Leeds.

Sharpe, S. (1977) *Just Like a Girl*. Penguin, Harmondsworth.

Shields, R. (ed.) (1992) *Lifestyle Shopping*. Routledge, London.

Smith, D.M. (1981) New movements in the sociology of youth: a critique. *British Journal of Sociology* 32, 239–251.

Smith, J. (1987) Women at play: gender, the life-cycle and leisure. In: Home, J., Jary, D. and Tomlinson, A. (eds) *Sport, Leisure and Social Relations*. Routledge, London.

Somnez, S. Shinew, K. Marchese, L. Veldkamp, C. and Burnett, G.W. (1993) Leisure corrupted: an artists' portrait of leisure in a changing society. *Leisure Studies* 12, 266–276.

Spruijt, E. and de Goede, M. (1995) Changing family structures and adolescent well-being. Paper presented at the Second European Sociological Association Conference, Budapest.

Stanley, L. (1977) Sex, gender and the sociology of leisure. In: Smith, M.A. (ed.) *Leisure and the Urban Society*. Leisure Studies Association, Manchester.

Stebbins, R.A. (1992) *Amateurs, Professionals and Serious Leisure*. McGill-Queens University Press, Montreal.

Stebbins, R.A. (1998) *After Work: The Search for an Optimal Leisure Lifestyle*. Temeron Books, Calgary.

Steger, B. (1996) Hurried work, hurried leisure and time to sleep: the case of Japan. Paper presented to the Conference on New Strategies for Everyday Life, Tilburg.

Stockdale, J. (1986) *What is Leisure?* Economic and Social Research Council/Sports Council, London.

Stockdale, J.F., Wells, A.J. and Rall, M. (1996) Participation in free time activities: a comparison of London and New York. *Leisure Studies* 15, 1–16.

Stokes, G. (1983) Work, leisure and unemployment. *Leisure Studies* 2, 269–286.

Streather, J. (1979) One-parent families and leisure. In: Strelitz, Z. (ed.) *Leisure and Family Diversity*. Leisure Studies Association, London.
Street, J. (1993) Global culture, local politics. *Leisure Studies* 12, 191–201.
Strinati, D. (1995) *An Introduction to Theories of Popular Culture*. Routledge, London.
Sugden, J. and Bairner, A. (1986) Northern Ireland: the politics of leisure in a divided society. *Leisure Studies* 5, 341–352.
Sulkanen, P. (1997) Introduction: the new consumer society – rethinking the social bond. In: Sulkanen, P., Holmwood, J., Radner, H. and Schulze, G. (eds) *Constructing the New Consumer Society*. Macmillan, London.
Sullivan, O. (1996) Time co-ordination, the domestic division of labour and affective relations: time use and the enjoyment of activities within couples. *Sociology* 30, 79–100.
Talbot, M. (1979) *Women and Leisure*. Social Science Research Council/Sports Council, London.
Talbot, M. (1990) Being herself through sport. In: Long, J. (ed.) *Leisure, Health and Well-Being*. Leisure Studies Association Conference Papers 44, Brighton Polytechnic.
Taylor, P. (1992) Commercial leisure: exploiting consumer preference. In: Sugden, J. and Knox, C. (eds) *Leisure in the 1990s*. Leisure Studies Association, Eastbourne.
Taylor-Goodby, P. (1985) Personal consumption and gender. *Sociology* 19, 273–284.
Taylor-Goodby P. (1986) Women, work, money and marriage. *Sociology* 20, 93–94.
Thompson, E.P. (1967) Time, work discipline and industrial capitalism. *Past and Present* 39, 60.
Thompson, S.M. (1990) Thank the ladies for the plates: the incorporation of women into sport. *Leisure Studies* 9, 135–143.
Thornton, S. (1995) *Club Cultures: Music, Media and Subcultural Capital*. Polity Press, Cambridge.
Thrift, N. (1989) Images of social change. In: Hamnett, C. McDowell, L. and Sarre, P. (eds) *The Changing Social Structure*. Sage, London.
Tinsley, H.E.A. Colbs, S.L., Teaff, J.D. and Kaufman, N. (1987) The relationship of age, gender, health and economic status to the psychological benefits older persons report from participation in leisure activities. *Leisure Sciences* 9, 53–65.
Tokarski, W. (1991) Research note: leisure lifestyle careers in old age. *Leisure Studies* 10, 79–81.
Tomlinson, A. (1979) Leisure, the family and the woman's role: observations on personal accounts. In: Strelitz, Z. (ed.) *Leisure and Family Diversity*. Leisure Studies Association Conference Papers 9, London.
Tomlinson, A. (ed.) (1990) *Consumption, Identity and Style*. Routledge, London.
Tomlinson, M. and Walton, D. (1986) A sporting chance. *Leisure Management* 6(5), 41–42.
Turner, R.H. (1964) *The Social Context of Ambition*. Chandler, San Francisco.
Tyler, M. and Abbott, P. (1998) Chocs away: weight watching in the contemporary airline industry. *Sociology* 32, 433–450.
Tyrell, B. (1995) Time in our lives: facts and analysis in the 90s. *Demos Quarterly* 5, 23–25.
Urry, J. (1990) *The Tourist Gaze*. Sage, London.

Urry, J. (1995a) *Consuming Places.* Routledge, London.
Urry, J. (1995b) A middle class countryside? In: Butler, T. and Savage, M. (eds) *Social Change and the Middle Classes.* UCL Press, London.
Veal, A.J. (1989) Leisure and life-style: a pluralist framework for analysis. *Leisure Studies* 8, 141–153.
Veal, A.J. (1993) The concept of lifestyle: a review. *Leisure Studies* 12, 233–252.
Veblen, T. (1925) *The Theory of the Leisure Class.* Allen and Unwin, London.
Vester, H-G. (1987) Adventure as a form of leisure. *Leisure Sciences* 6, 237–249.
Wait, P. (1996) Social stratification and housing mobility. *Sociology* 30, 533–550.
Walby, S. (1988) Gender, politics and social theory. *Sociology* 22, 215–232.
Wallace, C. and Kovatcheva, S. (1998) *Youth in Society: The Construction and Deconstruction of Youth in East and West Europe.* Macmillan, Basingstoke.
Walter, T. (1985) *Hope on the Dole.* SPCK, London.
Walton, J.K. (1977) Holidays and the discipline of industrial labour. In: Smith, M.A. (ed.) *Leisure and the Urban Society.* Leisure Studies Association, Manchester.
Walton, P. (1996) Enhancing positive attitudes to men working flexible and reduced hours. Paper presented to the Conference on New Strategies for Everyday Life, Tilburg.
Walvin, J. (1978) *Leisure and Society, 1830–1950.* Longman, London.
Wang, Ning (1996) Logos-modernity, Eros-modernity, and leisure. *Leisure Studies* 15, 121–135.
Warde, A. (1994) Consumption, identity formation and uncertainty. *Sociology* 28, 877–898.
Warde, A. (1995) Cultural change and class differentiation: distinction and taste in the British middle class, 1968–88. In: Roberts, K. (ed.) *Leisure and Social Stratification.* Leisure Studies Association Publication 53, Eastbourne.
Warde, A. and Hetherington, K. (1993) A changing domestic division of labour? Issues of measurement and interpretation. *Work, Employment and Society* 7, 23–45.
Wearing, B. (1993) The family that plays together stays together: or does it? Leisure and mothers. *World Leisure and Recreation* 35(3), 25–29.
Wearing, B. (1995) Leisure and resistance in an ageing society. *Leisure Studies* 14, 263–279.
Wearing, B. and Wearing, S. (1988) All in a day's leisure: gender and the concept of leisure. *Leisure Studies* 7, 111–123.
Wearing, B. and Wearing, S. (1992) Identity and the commodification of leisure. *Leisure Studies* 11, 3–18.
Wernick, A. (1991) *Promotional Culture.* Sage, London.
Whannel, G. (1986) The unholy alliance: notes on television and the remaking of British sport 1965–85. *Leisure Studies* 5, 129–145.
Wilders, M.G. (1975) Some preliminary observations on the sociology of the public house. In: Parker, S. *et al.* (eds) *Sport and Leisure in Contemporary Society.* Leisure Studies Association, London.
Wilenski, H.I. (1963) The uneven distribution of leisure: the impact of economic growth on free time. In: Smigel, E.O. (ed.) *Work and Leisure.* College and University Press, New Haven.
Williams, R. (1963) *Culture and Society 1780–1950.* Penguin, Harmondsworth.
Willis, P. (1977) *Learning to Labour.* Saxon House, Farnborough.
Willis, P. (1990) *Common Culture.* Open University Press, Milton Keynes.
Willis, P. Bekem, A., Ellis, T. and Whitt, D. (1988) *The Youth Review: Social*

Conditions of Young People in Wolverhampton. Avebury, Aldershot.
Wilson, J. (1988) *Politics and Leisure*. Unwin Hyman, London.
Wright, D. (1994) Boys' thoughts and talk about sex in a working class locality of Glasgow. *Sociological Review* 42, 703–737.
Wright-Mills, C. (1956) *White-Collar*. Galaxy, New York.
Wynne, D. (1990) Leisure, lifestyle and the construction of social position. *Leisure Studies* 9, 21–34.
Wynne, D. (1998) *Leisure, Lifestyle and the New Middle Class*. Routledge, London.
Yorganci, I. (1993) Preliminary findings from a survey of gender relationships and sexual harassment in sport. In: Brackenridge, C. (ed.) *Body Matters*. Leisure Studies Association, Eastbourne.
Young, M. and Schuller, T. (1991) *Life After Work*. Harper Collins, London.
Young, M. and Willmott, P. (1973) *The Symmetrical Family*. Routledge, London.
Yule, J. (1997a) Engendered ideologies and leisure policy in the UK. Part 1: gender ideologies. *Leisure Studies* 16, 61–84.
Yule, J. (1997b) Engendered ideologies and leisure policy in the UK. Part 2: professional ideologies. *Leisure Studies* 16, 139–154.
Zuzanek, J. (1996) Time pressure, stress and leisure: lifestyle and health issues facing industrial nations in the 1990s. In: *Newsletter, International Committee on the Sociology of Leisure*, July. International Sociological Association, Joondalup, Western Australia.
Zuzanek J. and Mannell, R. (1983) Work–leisure relationships from a social psychological perspective. *Leisure Studies* 2, 327–344.
Zuzanek, J. Beckers, T. and Peters, P. (1998) The harried leisure class revisited: a cross-national and longitudinal perspective. Dutch and Canadian trends in the use of time: from the 1970s to the 1990s. *Leisure Studies* 17, 1–19.

Index